SCIENCE,
TECHNOLOGY,
AND
CULTURE

SCIENCE, TECHNOLOGY, AND CULTURE

edited by
Henry John Steffens
and
H.N. Muller, III
The University of Vermont

AMS Press, Inc.
New York

All rights reserved. Published in the United States by AMS
Press, Inc., 56 East 13th Street, New York, New York 10003.

Library of Congress Cataloging in Publication Data

Science, technology, and culture.

   Papers and commentaries presented at four meetings between members of the faculty of the University of Vermont
and the executive staff of the Western Electric Co., held in
1972.
   1. Science—Social aspects—Congresses. 2. Technology—
Social aspects—Congresses. I. Steffens, Henry John, ed.
II. Muller, H.N., ed.
Q175.4.S37   301.24'3   74-580
ISBN 0-404-11275-7

Manufactured in the United States of America

TABLE OF CONTENTS

SCIENCE, TECHNOLOGY, AND CULTURE

Science as a Creative Art

Science and Social Responsibility

Afterword

PREFACE

In the spring of 1972, the Western Electric Company and the University of Vermont jointly sponsored four symposia held on the campus of the University of Vermont. The purpose of the symposia was to reexamine the differences between science and technology and to investigate the role of science in modern culture. The twelve participants in the symposia, who are either teachers in the academic community or are executives of Western Electric, were selected to reopen what was felt to be a long neglected dialogue between private industry and the university.

The Western Electric Company was convinced that the community of interest it shared—the improvement of society—far overshadowed the differences between it and the academic community. For this reason, Western Electric has co-sponsored, with other units of the Bell System, a number of symposia at Princeton University and the Universities of Kansas, Maryland, Michigan, and Vermont. The unity of the theme of the symposia at the University of Vermont distinguished those meetings from the others, and it was agreed that the material presented at the Burlington meetings should be revised and edited for publication.

Like each symposium, the four sections of this book have been planned to complement one another. It is, for example, important to reexamine the issues raised in "Science vs. Scientific Technology" before assessing the question, "Are There Two Cultures?" These first two chapters also provide the context in which to consider the possibility of "Science as a Creative Art." Finally, the issues in "Science and Social Responsibility" lead the reader to confront the cultural implications of science, technology, and the arts. Although each chapter of this book stands alone as a discussion of an interesting theme, it is hoped that the reader will find relationships among the four themes.

To provide a further opportunity to extend exploration of some of the major themes of the book, an "Afterword" has been written, and short bibliographies have been prepared as guides to easily available books on similar topics. Brief introductions to the four main sections of the book have also been prepared.

The success of the original symposia and, we hope, the usefulness of this volume are in no small part a measure of the enthusiasm of Donald K. Procknow, President, Philip A. Hogin, Executive Vice-President, and many other people in the Western Electric Company who found time in their busy schedules to lend their experience and thinking to the original sessions and who often remained in Burlington to talk with faculty and students. Generous encouragement from President Edward C. Andrews, Jr., Dean John G. Weiger of the College of Arts and Science, and many others at the University of Vermont is greatly appreciated. Finally, the editors and all of the contributors to this volume are especially indebted to Norman J. Rubin who, as Educational Relations Manager of the Western Electric Company, initiated and developed the format for the symposia at Burlington and provided useful editorial guidance in the preparation of this book.

SCIENCE
VS.
SCIENTIFIC
TECHNOLOGY

INTRODUCTION

The distinctions between science and technology have been obscured by the events of the twentieth century. Despite a clear historical delineation between them, the traditions of science and technology have become synonymous in contemporary usage. The confusion and lack of understanding of the differences between these traditions has serious implications for the future of science, technology, and society. L. Pearce Williams attempts to define this distinction in the first paper of this volume.

Professor Williams uses historical perspective in his arguments to distinguish between science and scientific technology. He chooses the terms "paradigm" and "problem-solving" from Thomas S. Kuhn's book, *The Structure of Scientific Revolutions*, and illustrates his use of these terms by suggesting that an overwhelming preponderance ("99.9%") of both "pure" scientists and scientific technologists are engaged in the same endeavor—problem-solving. He argues that if we shortsightedly concentrate our attention upon present experience with scientific technology, we might well draw the conclusion that both science and technology are fundamentally problem-solving activities. However, by casting our glance over the whole sweep of the history of the development of science, we discover that the real scientists are those rare creative souls—that one tenth of one percent of the people—who have worked with the "creation of a new world view." These are the real scientists who have taken themselves outside the methodological context of problem-solving and have succeeded in creating a scientific revolution.

Professor Williams stresses the uniqueness of the great innovators in the sciences like Johann Kepler, Michael Faraday, and Albert Einstein. He suggests that men of their genius were interested in the "form of the knowable world." There was a motivational impetus which distinguished them from others and an extra-logical component of their work which put them in the category of great creative artists. Such men's views "encompass the cosmos," and they are sensitive to the interrelationship between the various components which constitute the world. Viewed in this way, an unbridgeable gap can be seen between the genius and the problem-solver. The man who is intent upon a particular task is seldom concerned with all of the ramifications of his work.

Maintaining the distinction between science and scientific technology in this form has several implications for current practices in education and in the social consequences of technology. Michael Faraday urged men to "let the imagination soar." Professor Williams suggests that our science education programs today do not provide enough subject matter and exercise to enable the young scientist's imagination to soar. The current stress upon problem-solving has resulted in the insistence upon specific answers and has led to the gradual elimination of theology, literature, history, philosophy, and aesthetics as unnecessary, and even irrelevant distractions. In looking to the structure of education as both a culprit in the problems of modern society and the logical starting point in any systematic attempt to ameliorate those problems, Professor Williams initiates a theme that is to be reiterated throughout this book.

The modern stress upon problem-solving overshadows a man's quiet contemplation of the world. The problem-solvers must perpetuate their endeavors by finding answers to problems. Their justification for activity is the solution of a special problem, not the contemplation of nature. Within their narrow fields, the problem-solvers attempt to bend nature to their will by solving every problem nature presents. It is now clear, however, that they have created, and continue to create, more problems than we may be able to solve. For the first time in the three hundred year development of science and scientific technology, we are confronted with the question of being able to survive the problems created by the problem-solvers. Although Professor Williams' argument infers that society desperately needs to invest its resources in those things that could promote science (or creativity) and the discovery of new and broader world-views, even those in their commentaries who took the most optimistic view of the potential of technology insist upon the proportions and immediacy of the crisis. Therefore, in making such a clear distinction, this paper to a large extent defines the terms for much in the subsequent papers and commentaries, which continue to concentrate on the capacity of science or technology to grapple successfully with the gargantuan problems of modern society.

The Williams paper draws a wide range of commentary. Christopher Allen, an organic chemist, chooses a line of argument which places greater emphasis upon the process in science and technology, rather than on those people doing the work. He places those rare creative individuals that Professor Williams chose for his illustration into the larger category of problem-solving in both science and technology. Professor Allen argues that the methodology of science or technology, accumulating data, and formulating consistent conclusions make everyone a problem-solver. He also suggests there are serious problems raised by discussing extra-rational motivation in the development of scientists. He concludes by expressing a less pessimistic attitude toward developments in scientific education.

After pointing out that "technology, in general, has always brought blessings followed by curses," Dr. Jerry Cassuto agrees with Professor Williams that problem-solvers are creating many future problems. Often what one age regarded as a significant technological advance—a solution to a particular problem—has become a dilemma in subsequent times. Thus, "problem" is a perjorative term for a set of facts that may not be a problem to another age.

Conceding that today's education of scientists is too narrow, Dr. Cassuto contends that even if it were broadened, 99.9 percent of the scientific community would still be included in the ranks of the problem-solvers. However, he adds that they will be better able to approach science with an overall perspective that is all to often lacking today. He also questions Professor Williams' implication that a change in the educational process could increase the frequency of genius in society.

Dr. Cassuto is less pessimistic than Professor Williams about the capability of the technologists, particularly those trained to exercise a broadened perspective and to deal with new problems. "Who," he asks, "is to say that the new problems that have arisen and will continue to arise . . . cannot be solved?" Other contributors to this book pose answers to this crucial question, and the divergence of their responses indicate the parameters of the debate.

Eugene Anderson, an engineering executive, continues Dr. Cassuto's line of comment by suggesting that since most men have limited capabilities, we should concentrate upon inspiring and teaching "good men to do better things." Mr. Anderson asserts that no one knows how to create genius in a man, and moreover, no one can stop a genius from doing what he must. The views of most men, even those creative geniuses who have expounded new paradigms, are rooted in time and space and the experience of previous generations. He claims that the works of scientific geniuses would have been discovered. He then concludes his commentary by suggesting that our present uneasiness may be caused by the fact that science and its practioners may have advanced faster than our philosophy, sociology, and theology.

Professor Williams responds extensively to these commentaries, challenging many of the assumptions presented and indicating the seriousness of the debate. Scientists and technologists, despite their lip service to the method, rarely work from the collection of data to the development of a conclusion, and men like Michael Faraday do not simply rely on the work of those who have preceded them. There is a significant difference between science and technology, he reasserts, and that difference is crucial.

SCIENCE VS. SCIENTIFIC TECHNOLOGY

L. Pearce Williams
Chairman, Department of History
Cornell University

A great deal of ink has been spilled in the attempt to spell out the differences between basic or "pure" science and its presumably impure relation, applied science. Most distinctions are based upon the moral dimensions of the two activities. Pure science is done in an atmosphere of casual indifference to practical application, whereas applied research clearly lusts after specific (and profitable) effects. There is an element of Puritanism here, with the "pure" scientist representing God's saints and the applied scientist clearly leagued with Satan. Presumably, it is this theological background that lends such heat to the controversy.

I should like to return to this metaphor later, but at the start it would be well to take a different tack to try to see the problem more clearly. Let us first detail those aspects of "pure" science and scientific technology that are similar. We will be helped here if we use some of the terminology introduced by Thomas S. Kuhn in his work on the structure of scientific revolutions. Two terms are fundamental: paradigm and problem-solving. Let us leave paradigm aside for the

moment and concentrate on the latter term. Briefly, problem-solving is an activity similar to what, in military terms, is
referred to as "mopping up the battlefield." When a scientific
breakthrough occurs, new problems emerge or old problems
take on a new look. A whole host of investigators then turn
to the solution of these problems. This, surely, is a fundamental activity in both "pure" science and in scientific
technology. In both cases, what one is looking for is known
in broad terms before one begins. The "pure" scientist
expects to fit the results of his measurements, observations,
and calculations within the framework of the new theory
that has occasioned his investigation in the first place. The
scientific technologist expects his investigation to provide
him with precise data and operations whereby the effect
being studied may be described, reproduced, magnified, and
utilized. Both the "pure" scientist and the scientific technologist *do* much the same thing on the road to their
respective successes. Let me illustrate this with some examples. Modern nuclear theory makes various predictions about
the presence and arrangement of nucleons in the nucleus of
atoms. Most of the effort of High Energy physicists is
expended in making specific what was given in general terms
by the theory. Nuclear cross-sections provide the details
which fill in the general theory. Mapping cross-sections
involves great technical expertise and considerable mathematical sophistication. But, it is essentially a "mopping up"
operation. The "pure" High Energy physicist here does
exactly what the electrical engineer or scientific technologist
does when he investigates a potentially useful by-product of
"pure" scientific research. The history of the transistor
follows almost exactly the pattern I have described in nuclear
physics. The effect was announced and the scientific technologists set out to see what uses it might have. They mapped
the dimensions of the effect, made the various parameters
upon which the effect depended precise by means of this
mapping, and then channeled the results of their research
into application. They, too, used great technical expertise

and highly sophisticated mathematics. If one follows the behavioral scientists in insisting that human beings should be studied in terms of what men do, rather than in terms of the reasons given for doing it, then there is no clear distinction to be made between "pure" science and scientific technology in this area of problem-solving.

It is worth underlining the fact that problem-solving is what occupies about 99.9% of the members of the scientific community, whether engaged in pure or applied research, during their productive lives. Hence, it would appear only natural that we should conclude that the opposition expressed in the title of this paper is an artificial one. Consequently, I should proceed to write a brief conclusion. But my own Puritanism forbids me to stop here. Moreover, there are some loose ends that remain hanging out from this tidy package. There is that one tenth of one percent that I have not included among the problem-solvers; there is the paradigm left undefined, and there is the conflict between Satan and the angels that I barely mentioned. These items shall be discussed in the rest of this paper and will, I hope, justify its title.

The life of science, as we know it since the Scientific Revolution of the sixteenth and seventeenth centuries, does not consist merely of problem-solving. One of the values of history is that it allows the historian to escape the prison of contemporary impressions and detect large-scale movements that take more than one lifetime to mature. The present is filled with new applications of science and new solutions to old problems; it is not particularly rich in revolutionary new visions of the ultimate nature of reality. Thus, on the basis of contemporary science, it is possible to see science as *primarily* a method of problem-solving and thereby to insist upon the fundamental unity of "pure" science and scientific technology. But, if one casts one's gaze backward, the striking feature is the creation of new world-views that so fundamentally changed man's perceptions of the world in which he lived that all relations within that world had to be redefined. It is this feature of science that involves that one tenth of one percent and that literally makes all the difference in the world to the life of science.

This revolutionary change in world-view is what Kuhn sees as a change in paradigm. The paradigm dictates what a man sees around him; change the paradigm and he must learn to see again. That it involves a severe shock to those who must change paradigms is well illustrated by the reaction to the Copernican revolution. Far more was at stake at that time than a simple question of whether putting the earth or the sun at the center of the universe made more astronomical sense. The English poet, John Donne, saw what was involved and expressed it in eloquent terms:

> [The] new Philosophy calls all in doubt,
> The Element of fire is quite put out;
> The Sun is lost, and th'earth, and no man's wit
> Can well direct him where to look for it.
> And freely men confess that this world's spent,
> When in the Planets, and the Firmament
> They seek so many new; then see that this
> Is crumbled out again to his Atomies.
> 'Tis all in pieces, all coherence gone;
> All just supply, and all Relation:
> Prince, Subject, Father, Son, are things forgot,
> For every man alone thinks he hath got
> To be a Phoenix, and that then can be
> None of that kind, of which he is, but he.

The real question we want to get at here is: why do paradigms change? One answer, to which many if not most, practising scientists would subscribe is that the normal activity of problem-solving gradually builds up a body of factual evidence which will not fit within the current paradigm. When, so to speak, this body of fact reaches some kind of critical mass, the old paradigm is ruptured and a new one (almost automatically) takes its place. This view would minimize the importance of my one tenth of one percent. They would be merely the lucky ones, on the spot when the paradigm broke and, therefore, able to put a new paradigm together from the wreckage of the old. If we accept this interpretation, then once again we must admit that there is little difference between the "pure" scientists and the scientific technologist. There is evidence, however, that this is not an acceptable view.

Let us look more closely at the sixteenth century when the Copernican hypothesis was competing with the older Aristotelian and Ptolemaic paradigms. Two men are outstanding in this period—Tycho Brahe and Johann Kepler. I would argue that Brahe is a perfect example of the problem-solver, whereas Kepler is one of my one tenth of one percent. A comparison of their contributions to the development of astronomy will prove instructive.

There were a number of difficulties in the way of accepting Copernicus' suggestion that the earth and the other planets went around the sun. Not the least of these was the problem in physics raised by the earth's annual revolution. Why should the earth revolve around the sun? The fact that a large number of hitherto unrelated astronomical facts could be made to make sense if the earth did revolve around the sun was hardly a persuasive reason for a sixteenth century astronomer (or layman) to fly in the face of common sense, experience, and tradition. Tycho Brahe took this common-sense view and preserved a geocentric universe. What he drew from Copernicus was the sense of dissatisfaction with the old cosmologies and the stimulus to define astronomical parameters as precisely as possible. His problem, simply stated, was to discover the *actual* positions of the heavenly bodies. He did a cosmic cross-section, if you will. In the process, he destroyed the Aristotelian paradigm. Problem-solvers, it can be admitted, do bring paradigms down.

But, it was not Tycho who provided the new paradigm. It was his assistant, Johann Kepler. It was Kepler who broke fundamentally with the Aristotelian-Ptolemaic world-view and who ushered in a new era in physics and cosmology. The point is worth insisting upon for Kepler is usually treated as the man who established Copernicanism on a sound factual base. This view is fundamentally incorrect, for the only Copernican element in Kepler's vision is heliostaticity. The rest is Kepler's vision, and it is of such originality that it took more than a generation for it to be accepted and taken seriously.

Copernicus had tried to remain within the old paradigm by preserving the circular motion of the planets as well as retaining the crystalline spheres in which the planets were assumed to be embedded. The earth turned around the sun because the crystalline sphere forming the terrestrial orb turned. This orb turned because, as Copernicus pointed out rather feebly, it is the nature of a sphere to turn upon its axis! By accurate observations, Tycho showed that comets cut across the planetary orbits without encountering crystalline spheres in their paths. Such spheres, then, simply could not exist and the planets hung, mysteriously, in empty space. Why they moved as they did did not seriously concern Tycho; it did seriously concern Kepler.

At this point, it is no longer possible merely to follow the logic of astronomical development. Such a logic dictated something like the following: Copernicus had shown that the Aristotelian-Ptolemaic system could be made more intellectually economical if the center of the cosmos were moved from the earth to the sun. Tycho's observations removed the physical substratum of the Aristotelian paradigm, leaving an enormous hole in the physics of celestial motion. The logical thing to do, then, was simply to complete Tycho's work by mapping planetary orbits, and forget the physical problem. That is what Kepler did (at first). He found a solution that would have satisfied 99.9% of his fellow astronomers. The orbit of Mars, he found from Tycho's data, fit an oval orbit to within eight minutes of arc. For the above-mentioned 99.9%, that would have solved the problem. Kepler was tempted to stop there, too, but he resisted the temptation.

Why? According to Kepler himself, God had given us Tycho Brahe who could observe accurately to within four minutes of arc, and it would be sinful to rest content with an orbit that was off by eight minutes. For the problem-solver, eight minutes could have been construed as four minutes on either side of complete accuracy and, therefore, within the observational limits of Tycho's accuracy. But for Kepler, the real difficulty lay with his oval. It was simply not an elegant enough mathematical figure to satisfy him. Since his first introduction to astronomy, Kepler had been intoxicated by the prospect of discovering the mathematical harmonies of God's creation. It should be noted that there is no compelling logical reason for believing that God is primarily a Geometer. We are reminded of the Scriptural remark that God works in many and wondrous ways. To restrict Him to the mathematical mode is, to say the least, idiosyncratic. Yet Kepler would have it so. The Copernican system appealed to him because of its apparent mathematical superiority over the Ptolemaic. His earliest discovery that the planetary orbits could be inscribed in the five Platonic regular solids sent him into ecstasy. That the God who could accomplish all this would be content to permit a planet to move in so lowly and insignificant a curve as an oval was unthinkable. And so, Kepler persevered and found, to his delight, that the real curve was an ellipse. Now *that* was the kind of curve that Geometer God would choose and Kepler could rest content with his discovery. But, of course, it made the physics of the situation impossible. It had been difficult to explain why planets moved in circles; no one could explain why a planet would move in an ellipse. No one, that is, before Newton. Newton's solution to the problem of accounting for the motion of the planets in elliptical orbits created classical dynamics. But that required the formulation of the concepts of central forces, universal gravitational attraction and inertial motion. Kepler's new astronomy required a new physics.

Again a few points are worth making explicit. First, the "fit" between Kepler's data and the ellipse was good, but not perfect. The orbit of Mars (or of any planet) is not, in fact, an ellipse but a complicated path caused by perturbations of the other planets. Kepler saw an ellipse, at least in part, because he wanted to see one, not because the ellipse was so obstrusive that it simply had to be seen. Secondly, Kepler appeared able to take the revolutionary step of ascribing an elliptical orbit to a planet with complete equanimity. The size of the step should be noted. Ever since the Pythagoreans it had been argued that the heavenly bodies *must* move in perfect circles. Plato and Aristotle had devised rather clever and complicated arguments to prove that heavenly motion was circular. By the sixteenth century, no one doubted that heavenly motion was circular. Even Galileo, certainly no conservative in other areas of physics and astronomy, could never free himself from belief in the necessity of heavenly circular motion. Yet Kepler could and did insist that heavenly motion was elliptical. He was the sole believer in that truth for a generation. Clearly there is something more here than simple problem-solving, and it is not difficult now to discover what it is. Rather subtly the problem has shifted from that of accounting for the motion of the planets to accounting for Johann Kepler. Why was it he, alone among all his contemporaries, who could see and enunciate Kepler's First Law? It was not merely a case of having a monopoly of the data, as is proved by Galileo's refusal to accept Kepler's findings. It obviously has to do with the psychic and inner needs of Johann Kepler himself. Neither time nor my competence permits any serious foray into the depths of Kepler's soul. Arthur Koestler has tried his hand in *The Watershed,* and I refer you to this work. According to Koestler, Kepler felt it necessary to impose an order upon the phenomena of nature if he were to retain his sanity. With a slattern for a mother, a mercenary soldier for a father, almost total chaos for an environment, and a talent for mathematics, it does not seem far-fetched to suggest that Kepler created his own vision of the cosmos and that that vision was one of exact and ordered mathematics. The problem that Kepler solved was the intensely personal one of deriving order from chaos through the instrument of his own mind. His mathematical genius and obvious respect for empirical data were part of the solution, for without them he could have seen nothing. But the primary element was the personal equation. It was this that drove him to continue his calculations beyond the oval; it was this that inspired him to continue to look for those secrets of the universe that he knew must exist given his faith in a Geometer God.

I have examined Kepler in some detail, because I am convinced that he is not unique in the history of science, but is a fair representative of that one tenth of one percent who drive science onward over broad fronts. With Kepler in mind, it is possible, perhaps, merely to mention two other creative scientists who resemble him: Michael Faraday and Albert Einstein. Faraday is interesting in our context because he was not a mathematician. If I am to forestall the objection that Kepler's dissatisfaction with an oval was merely the response of anyone imbued with mathematical learning, then it is important to show that the same kind of refusal to accept the easy, but personally distasteful, answer to a problem is possible among non-mathematicians. The case is clearly illustrated by Faraday. Almost from the beginning of his scientific career, Faraday felt it necessary to believe in the essential unity of the forces of nature. There was no good scientific reason for such a belief, but there was a sanction for it to be found both in metaphysics and in Faraday's theology. One consequence of this belief was that the forces of nature should have mutual effects on one another. Thus, for example, electrical forces should do *something* to light. Faraday sought the effect for over thirty years and never found it. Yet his faith in its existence was never shaken and, in 1875, after his death, it was discovered. Faraday, the careful experimenter, the man who publicly, time and time again, insisted upon the importance of staying within empirical evidence, staunchly and steadfastly held to the belief in an effect that he could not detect. Note again that this position is quite different from that of the problem-solver. Although Faraday's experiments might be interpreted to mean that the effect existed, failure to detect it, according to the empirical canon of the problem-solver, should have discredited it. Faraday, however, was driven by more than evidence; he had painfully constructed a new world-view—the view of field theory—in which the doctrine of the unity of force was of fundamental importance. He could no more abandon his belief in the influence of electricity upon light than Kepler could give up his belief in the essential mathematical nature of the world. Like Kepler's, his view derived essentially from a theology and had about it the characteristics of religion, not that of experimental problem-solving.

There is a difference between Faraday and Kepler that may be used to emphasize my point about the uniqueness of viewpoint of the great innovators in science. Whereas it could be argued that Kepler was led to his formulation of the First Law by his sole possession of the facts upon which that law was based, it is quite otherwise with Faraday. Faraday inherited a mature paradigm of the nature of electricity and magnetism. Because, in his opinion, this paradigm led to atheism and materialism, Faraday rejected it and, *using exactly the same facts as were available to his contemporaries*, created field theory. No one knew what he was about until after his death. What Faraday saw, he saw with an inner eye and the view was visible to him alone. The secret of his vision was his intense concern to leave a place for God in the universe from which the French physicists had expelled Him.

Albert Einstein resembles both Kepler and Faraday. Like Kepler, Einstein thought God must be a Geometer. What distinguishes him from Kepler was that he was an atheist! As it did for Faraday, the field served as the unifying thread for this *uni*verse. It used to be a commonplace that Einstein had merely added the final step to a logical chain that had its first link in Newton's definition of absolute space and time, and penultimate link in the Michelson-Morley experiment. It is now fairly well established that Einstein's 1905 paper on the motion of electrodynamic bodies owed its origin to a youthful obsession. In his autobiography, Einstein relates how, at the age of sixteen, he wondered what the universe would look like if he were riding the crest of an electro-magnetic wave. Again, if there is a problem to be solved here, it is not one immediately suggested by the paradigm of classical field theory. Rather, it is another personal equation, the roots of which are found deep in Einstein's personal development. One root is the aesthetic sense which appears in the rejection of the asymmetry to be found in the classical field equations. Another, manifested particularly in his writings on the General Theory of Relativity, is Einstein's deep-seated belief in a deterministic and harmonious cosmos. The mathematics is almost infinitely more sophisticated and the view even more cosmic, but the basic framework is one that Kepler would have appreciated. This belief is what distinguished Einstein from his contemporaries during the late twenties and thirties when quantum mechanics was being created by Bohr, Heisenberg and Schrödinger. It is usual among problem-solvers to accept the method that solves the problem. This is precisely what Einstein could never do. He agonized for years over the principles of quantum mechanics, convinced that the generally accepted probabilistic interpretation *had* to be wrong. The answers, however, were correct; the problems were solved. And Einstein went his lonely way to the day of his death. Ultimately, he could appeal only to his strongly felt inner sense that "God does not play dice with the cosmos." This is hardly a methodological principle of universal acceptability!

What are we to make of all this? Are Kepler, Faraday and Einstein merely scientific "sports" who did good science in spite of their obvious intellectual eccentricities? Or, rather, do they tell us something of fundamental importance about the life of science? It is possible to argue both positions and impossible to prove either. Nevertheless, I shall suggest that the latter is preferable. It seems to me that it is possible to discern two processes within the evolution of modern

science. The problem-solver is both ubiquitous and important. And, since we have seen that problem-solving does not
differ significantly in "pure" science and/or in scientific
technology, it is important to note that science and scientific
technology are indistinguishable on this level. But, the
Keplers, Faradays and Einsteins are not problem-solvers. The
one characteristic of their scientific work which unites them
is its universality. Although a twitch of a needle or the
swinging of a compass point may stimulate their researches,
their views encompass the cosmos. They are not, basically,
interested in singularities but in the form of the knowable
world. They are the grand strategists of science. Their work
creates the battlefield which later generations mop up.
Without them, there would be no problems for the problem-
solvers to solve. And, in this sense, they break through the
vaunted democracy of the republic of science. They are the
aristocrats of science, for upon their insights and visions
depends the vitality of the scientific enterprise. Problem-
solvers are plentiful and almost interchangeable. One *can* say
of them that if one had not discovered or calculated something, someone else would have. But, a Kepler, a Faraday or
an Einstein, like a Beethoven, is unique. That element of
extra-logic that characterizes their work puts them into the
category of the great creative artist. No one would suggest
that if Beethoven had never been born, his Kreutzer Sonata
would have been written by someone else. Neither would the
works of my one tenth of one percent.

If my analysis is correct, an unbridgeable gulf exists
between this "pure" science as epitomized by my one tenth
of one percent and the problem-solvers, whether they be
"pure" scientists or scientific technologists. Does this distinction, however, have any meaning beyond the simple intellectual recognition that such a distinction exists? I believe it
does. It is appropriate to bring it out clearly in an educational
institution for it seems to me that it is precisely in the area of
higher education in science that failure to recognize this
distinction has done almost irreparable harm to the growth of

science. Modern education in science is geared almost entirely to the problem-solver. From the moment of his appearance on the university campus, to the day when he departs with Ph.D. in hand, the embryonic scientist is faced with problem-solving as though it were the totality of science. Problems are set, the "right" answer is expected, and grades are determined by the facility with which the right answer is forthcoming. The result of this education, both of the "pure" scientist and the scientific technologist is to force a student's mind into a peculiar set which insists upon isolating a "problem," investigating it within as narrow a compass as possible, and providing an "answer." Perhaps that is all that education can do, and perhaps it is churlish of me to suggest that the result is potentially stultifying. Yet, looking around at contemporary science, one is struck by the essential dullness of most of what is being done. The problem-solvers are pouring solutions in ever-widening streams into the technical journals, but it all has a plainness and a sameness that belies the claims to novelty and excitement offered by the salesmen of science. I suspect my one tenth of one percent is being drowned, or at least drowned out. The mark of this group is its apprehension of the wholeness, of the totality, of the world. They live by visions, seizing upon all areas of human experience to create their own view of what is real. To be sure, they must bring their vision within the framework of science and express it in the austere language that science demands. But the sources of the vision are manifold. "Let the imagination soar," Faraday once wrote, but where in our modern education is there any food upon which the imagination may feed? Theology, literature, history, philosophy, aesthetics—all have been carefully weeded out of the curriculum as being distractions or irrelevant to the main task of solving that next problem. And yet—heresy of heresies—may it not be that the imagination, like the reason, thrives on exercise and that the flight of fancy may be of equal importance to the growth of science as the computer program?

I conclude with a final thought that may also be important in distinguishing the two kinds of science which I have considered. When reading Kepler, Faraday or Einstein, one is struck by their reverence for the world they are attempting to understand. When one sees something whole, there is a certain value to its wholeness. Such a wholeness, at least for Kepler, Faraday and even Einstein, was an attribute of deity, and the harmony of the cosmic parts was testimony to the refinement of the Creator. Each of the men I have mentioned was aware of the practical applications that could be made of their ideas, but each, more or less, disdained to make them. Each, I suspect, would have claimed that "pure" science consisted precisely in the contemplation of the cosmic whole. Application was, in a sense, a human presumption upon the Divine. That the fruits of science could raise man's estate, each acknowledged; but each, it should be noted, was something of an ascetic. The vision sufficed. Not so with the problem-solver. What, after all, is the real difference between a problem in nuclear physics and a problem in radio communication or in applied chemistry? Problems exist to be solved. It was not always thus, and perhaps we have come to the point where we can no longer afford so many problem-solvers. The problem-solvers are now the ones who are creating many of the problems for the future. Perhaps contemplation of Truth is, as Aristotle would have agreed, the *summum bonum*. Faust is a modern creation who is becoming increasingly embarrassing with age. If God had wished to make man the master of nature, he would have made him an angel. And so, with this final rhetorical flourish, I have returned to my original metaphor. The Saints see the whole and contemplate it; the rest would bend nature to their will by solving every problem nature presents. In so doing, they may well solve all problems but the ultimate one of how man himself is to survive.

COMMENTARY

Christopher Allen
Associate Professor of Chemistry
University of Vermont

A problem is, as a dictionary might say, a question posed for solution or consideration, something which leads us to approach the unknown. Within this definition, I want to reexamine Professor Williams' classification of science and scientific technology and include his "one tenth of one percent" within the larger group of problem-solvers. In the process, I want to consider only what is being done, not who is doing it.

The basic process of science is the observation of physical reality and the attempt to develop a one-to-one correspondence between some model system and observation. This model is then usually referred to as a description of nature. At some arbitrary start in the scientific process, pure observation (or data collection) occurs and leads to synthesis or model formulation. There then follows the continual process of model testing and refinement. A synthesis is produced or a concept is developed. Eventually, this synthesis or concept may prove unsatisfactory, and it is at this time that a scientific revolution, a change in world-view, occurs.

Where then do science, technology, and the one tenth of one percent fit into this scheme? First, technology is the observation (or development of methodology for observation) which is not directed towards model formulation or testing, and secondly, science is the formulation and testing of models by the analysis of observation. Professor Williams has suggested that the testing phase consists of "fitting observation to established theory." I believe that the process is more complex than this phrase indicates. In the ideal, at least, a preconceived "correct" answer should not exist. In fact, in the most intellectually satisfying work, the accepted model appears to be inadequate or incorrect. This ideal is not just a manifestation of pleasure in perversity, but rather it is based upon the fact that the formulation of scientific laws on the basis of models is basically an inductive process. Hence, a law can never be proven, but it can be disproven. Based upon induction, the scientific process involves establishing a crucial test which will allow the evaluation of the viability of the model.

What then is the function of the one tenth of one percent? I would say that their function is to declare that the emperor and his whole kingdom are naked. They must fit everyone with new clothes. Their gift of genius is to see that the whole system of models which comprises a world-view is not in need of further modification, but rather, it is in need of demise. They must start from scratch to build a new system. The motivation and inner life of such people, or at least some of them, may well fit the pattern suggested by Professor Williams. However, a problem in extra-rational motivation as it relates to science should be recognized. A person could possibly be driven to find an answer which does not exist or to develop an unsatisfactory model. For example, if Kepler had been deeply involved in primitive religions and had been overwhelmed by the appeal of the Earth Mother concept, he would then have been driven by that inner need to apply his mathematical genius to try to fit Tycho Brahe's data to a geocentric universe. In a less facetious way, we may view

Einstein's problem in a similar light. We can consider Einstein as grounded in the nineteenth century belief in the necessity of a deterministic universe. He could not free his mind to approach the problems raised by quantum theory objectively.

Basically, I have approached the separation of science and technology in terms of a specific process or in terms of what is done, not who does it. This approach is important because any one person may be involved in both science and technology. In fact, the one tenth of one percent may also be involved in both as exemplified by Michael Faraday. I am not saying that there is no difference between science and technology, but rather, that the labels of scientist and technologist are misleading.

As to education, the trend in teaching science and engineering over the past few years has been to allow the student a greater variety of intellectual experience rather than to eliminate systematically the use of non-technical materials. This freedom of intellectual experience is, I believe, basically a positive development. Moreover, I personally have enough faith in scientific genius not to worry about its being drowned by overly rigid educational approaches to the development of human intellect.

COMMENTARY

Jerry Cassuto
General Medical Director
Western Electric Company

In the history of my own medical profession, I cannot help but wonder about those who, over the years, might be regarded as paradigm-changers. Certainly, William Harvey's demonstration of the circulation of the blood, which replaced Claudius Galen's ebb-and-flow concept in the seventeenth century, was both courageous and revolutionary, as was his new experimental method in medicine. Rudolf Virchow's conception of the cell as the center of all pathological processes changed the direction of nineteenth century medicine. The magnificent contributions of Louis Pasteur, Sigmund Freud, and Paul Ehrlich also would indicate a scope of thought and operation above and beyond simple problem-solving in their respective fields.

But, as C. P. Snow points out, medicine (which in human terms may be the most significant of all technologies) presents a sharp example of the two-faced nature of technology. By its reduction of infantile mortality, a very positive good, it has also damned us with perhaps our greatest danger—a flood of population. By prolonging life, it has also caused more people to suffer the ravages of the chronic diseases and the loneliness and problems peculiar to old age. Technology, in general, has always brought blessings followed by curses. There seems to be a constant need to be in a race to keep up with the solutions to these initially unforeseen problems. Admittedly, as Professor Williams has stated, the problem-solvers are now the ones who are creating many of the problems for the future. But, at the same time, they have solved many critical problems of the past. Who then is to say that the new problems that have arisen and will continue to arise as a by-product of past and future solutions, cannot be solved?

Professor Williams, in his conclusion, states that "the Saints see the whole and contemplate it; the rest would bend nature to their will by solving all problems but the ultimate one of how man himself is to survive." But let's try to think of that statement in a different way. There are so many problems critical to man's survival that if we were to attempt to create more individuals who could see "the whole and contemplate it," would there then be a sufficient number of people applying themselves to the problems that must be solved? To quote Snow again: "The only weapon we have to oppose the bad effects of technology is technology itself." Can a change in the educational process truly increase the frequency with which the Keplers and the Einsteins come along? The uniqueness of these men is such that a new educational orientation would not significantly affect their numbers. Certainly today's education of scientists is much too narrow. More attention to the humanities is needed not only for the imagination, but for the growth of the individual as both a scientist and a member of society. As I indicated, we will still have 99.9% of the scientific community included in the ranks of the problem-solvers, but with a difference. They will be better able to approach science with an overall perspective that is all too often lacking under current conditions.

To conclude on an ironic note, it appears that the

problem-solvers working with DNA and genetic coding might eventually enable us to genetically manufacture genius on the order of an Einstein. Although there are dangers inherent in such manipulation, once again, as always, the problem-solvers may have created new problems for us to confront.

COMMENTARY

Eugene Anderson
Director of Engineering
Western Electric Company

I find that I differ with Professor Williams in about the way one might expect a pragmatic engineer to differ with an historian of science. In the first part of his paper, Professor Williams defines the vast majority of scientific activity as involving the methodology of the problem-solver. He then leads us adroitly to the conclusion that science and applied science are very much alike. In fact, he argues that there is a correlation factor of 99.9%—a figure certainly satisfactory enough for my engineering colleagues. At this point, Professor Williams states his thesis that true science is the paradigm or great leap forward brought about by men of genius like Kepler, Faraday, and Einstein. It is these men, he says, who are the true scientists. All others are the fillers-in of the theory, the refiners of the data, and the appliers of the new knowledge. Some of these latter types of men are good at their task, some not as good, but all combine to produce a tremendous volume of published material. On the other hand, if these "great men" had not lived, we might indeed be stumbling, awaiting their findings. Professor Williams also finds possible fault with science education and with the requirement of industry, which tend to favor the more mundane scientists and the applied scientist and mitigate against the genius.

I am not at all convinced that the scientific discoveries of these great men would not have been found in due course. They were indeed great men, geniuses beyond a doubt, obsessed with a concept and driven by the determination to understand. I have no argument that, by living and contributing to knowledge, they advanced our understanding sooner than might have been expected—by a year or a hundred years. But, each of these men were preceded by generations of searchers, with a predetermined pattern of nature, in existence, awaiting their discovery. If this is not so, then Kepler should have anticipated Newton, and Newton, in turn, all of Einstein. Moreover, history does not record the failures of men of equal conviction and equal determination. I would think that this is in sharp contrast to scholarly and aesthetic fields of art and literature, where a single genius such as Beethoven or Shakespeare or Michelangelo may indeed never be repeated nor anticipated. The works of these geniuses would not have been discovered in due course.

I do agree that much of teaching is undoubtedly mundane and much of the output of science is pedestrian, and both are part of us 99.9% problem-solvers. We must be taught, not as geniuses, but as we are, to do our work in the best way we know how, within our limited capabilities. Although I do know it is possible to inspire and teach good men to do better things, I do not know how to bring genius to a man. Moreover, I would suspect that no one knows how to stop a genius from doing what he must.

In the final part of his paper, Professor Williams gave us a twist which I would interpret as being the question of "just what is man up to?" Is he fulfilling his destiny by finding the truth of nature, or destroying himself by insisting on finding such truth and bending it to his will for purposes or reasons he chooses? This, I suspect, is the real dilemma he wished to

leave with us. If so, then who among all men is to recognize
the truth of truth and present it so that we may all perceive
it. As Professor Williams says, God did not make us angels,
but men.

We are left then, with the problem of what is man to do.
Should he pursue scientific knowledge for its own sake but
not apply it, or should he apply it as if he were some
God-like arbiter decreeing it as being best for society?
Indeed, we are continually faced with developments posing
this dilemma. Who would deny nuclear studies and their
"beneficial" values to society? Yet, all stand in horror that
man may literally destroy himself. There has been "great
progress" in medicine in saving and prolonging life, and we
may even now be standing on the threshold of creating life
itself. But this capability too raises questions—of how long
life should be prolonged and at what expense to society.
Should there be a eugenic selection? In fact, I am not at all
sure that doctors and lawyers would agree on a definition of
life itself or on a determination of precisely when a man is
alive and when he is dead. Here, I think, is the stone in
Professor Williams' snowball. Is he not suggesting that
science and its problem-solvers have advanced faster than our
philosophy, sociology, and perhaps theology can keep apace?
Is this, perhaps not the root of our uneasiness? The ultimate
question is: what to do about it, if anything.

RESPONSE

L. Pearce Williams

It delights me that Professor Allen chose as a jousting field the history of science, because his choice gives me a certain confidence in stating that he is wrong. He is wrong for an obvious reason, and it is the reason that scientists are usually wrong. He chose to consider an ideal case. There are no ideal cases in history; there are only the imperfect, real cases of what happened. What happened is simply not what he describes. I know of no examples of a major scientific theory in the history of science that resulted from data collection. What emerges from data collections have been fantastic collections of data. Scientists simply do not do science as an effort of perfectly collecting ideal data; nor I suspect does Professor Allen, because he is not an ideal chemist. He is a real chemist, and he does not sit and simply collect data. You collect butterflies, or you collect stamps, or you collect matchbox covers. The results of these collections are well known; they are known as hobbies. In dealing with science however, you need to think, and you do not think on the basis of collecting data. Indeed, I would argue that scientific theories tell you what facts to discover. The theories tell you what data is; you do not simply collect data. Until you have a scientific theory, there is no data but just a lump of something somewhere which you never previously noticed. The minute you have a theory, you know what the data is. Then you can go out and collect it, because it will make some sense.

Furthermore, I do not like Professor Allen's distinction between science and technology that technology does not involve itself with model formulation and science does. I am a former chemical engineer. For example, if you want to know what happens in the turbulent flow in a pipe, you spend a great deal of time trying to determine a model to give you some ability to create equations which will give you some insight into what is occurring in that pipe. The problem seems to me to be a technological one, or certainly a problem of technology which involves model formulation. This problem-solving may not go beyond the creation of a quasi-empirical formula; that is, it may not continue further to working with the ultimate nature of molecular forces that will be determined by the various conditions, and so on. I think simply that Professor Allen is wrong in his picture of science. His view is interesting, because it is very typical of what scientists say science is. I do not honestly believe that Professor Allen does scientific work in this way, and yet I have never heard a scientist who does not describe science in a similar, ideal way. It is a wonderful kind of false image, and I do not understand why it stays alive so long. It obviously has some use, other than being used in arguments against humanists.

One of Mr. Anderson's fundamental points is a beautifully metaphysical one, because we can never prove it. As Mr. Anderson presents it, the point is always assumed to be true, and I would at least like to try to shake his confidence in it. Mr. Anderson states that had Einstein, Faraday, and Kepler never lived, in ten years or so what they did would have become part of the knowledge of science. This is a metaphysical point, because they did in fact live, and therefore we can never know the answer to the question of what would have happened if they had never lived. However, I do feel that there is some interesting evidence to indicate that Mr. Anderson is wrong.

As I have indicated, there was no compelling reason for Michael Faraday to invent field theory. He invented field theory with precisely the same facts that were available to his contemporaries. We know today that an exact equivalent exists between particle and field theory; that is, any problem that can be solved by one theory can be solved by the other. Consequently, there is no necessity for field theory to ever have been developed in the history of science. It is perfectly possible for science to have developed in such a way that the concept of field never arose. In fact, every problem which can be solved with the concept of the field can be solved with the concept of the particle or the quasi-particle. If the road could have been absolutely straight from Marquis de Laplace's famous remark in the nineteenth century, one could go from Laplace to Hans Bethe in the twentieth century without having to digress into field theory. In his *Essay on Probability*, Laplace remarked that if an infinite mind knew the position and momentum of every particle in the universe, that mind could predict the exact course of that universe to eternity. Laplace, incidentally, felt that he did not qualify as the infinite mind. There is no inherent logic in the physics of the situation that demands the invention of field theory. Faraday should be considered somewhat unique; his invention of field theory is a rather interesting aberration in the history of science, brought about not by any compelling reason, but by theological reasons and political reasons. To believe Laplace was to believe in an atheistic, deterministic, materialistic universe. To believe in an atheistic, deterministic, materialistic universe was to believe in the ultimate dissolution of society. The road to that dissolution had been made abundantly clear in the French Revolution which had occurred in Faraday's youth. The distance between Laplace's particle physics and Robespierre and the Terror was, to Faraday, a very short one indeed. If you accept Laplace, the streets of London will run with blood. Perhaps I exaggerate, but nevertheless, the exaggeration does reveal an important motivation in Faraday's line of development.

We must consider this point that science is not automatic, because it has a very important consequence. What Mr. Anderson is really saying is that some day, in a personal rendering of Cole Porter's lyrics, "we'll live to see machines do it." This may happen if science is merely a case of the development of the inherent logic of situations. This is the implication of the remark that if Faraday hadn't done it, Joe Schultz would have. They are interchangeable, and if true, then machines ought to be able to be substituted into the scientific process. Indeed, the computer would be the scientist of the future. This line of reasoning is wrong. Computers can help, but science still needs human vision.

Finally, I am struck by Mr. Anderson's statement that science has forged ahead while the humanistic disciplines and perhaps the behavioral sciences such as sociology, anthropology, alchemy, and astrology have lagged behind. With that statement goes a certain degree of "Gee, too bad, fellows. You know we can't hold up. And, why are you so slow? Why don't you get moving and stop fiddling around with unimportant things and get these humanistic studies and morality up to where they should be."

I shall change the metaphor. I would like to reinstitute an old metaphor which dominated Western civilization until the eighteenth century. I do it with a certain amount of joy, for I happen to be an atheist. The metaphor deals with something known as the fruit of the tree of knowledge, good and evil. The implication in Eden was that in fact man can know too much. And, if he knows too much, he may destroy himself. It is that tradition to which I appeal in a very insidious way in my paper. It is that tradition in which I suggested that the contemplation of nature may, in the long run, be a more healthy activity than the manipulation of nature. The importance of this seems very clear. In the long run, contemplation may be superior to manipulation, because manipulation doesn't give us a long run.

We are living in a period when for the first time in three hundred years the good from science and technology is outweighed by evil. For three hundred years science and technology were supported by a basic belief which generally seems to be carried through in terms of practice: that the good effects of science and technology outweighed the evil effects. People were quite aware of the evil effects. The problem of pollution in Faraday's London was intense. It was Michael Faraday who devised the solution—and I use the term, solution, with a somewhat heavy heart—of protecting the paintings in the National Gallery by covering them with plates of glass. Although Londoners saw any number of beautiful reflections of themselves trying to see what was behind the glass, something had to be placed over the paintings, or they would have been blackened by the end of the nineteenth century. Faraday knew the problems of pollution and knew them well. We however are now living in a period when there is a real sense that what science does is no longer good in the balance; the good we get from science may no longer outweigh the evil.

I am not hostile to science, but I am trying to sound a warning to science and technology. I am trying to assert two things. First, scientists cannot do business as usual anymore. It is no longer possible simply to say, "Well, we have outstripped those other studies." Even if that kind of accomplishment is true, it is no longer sufficient to permit social approval of technologies continuing to outstrip other studies. Clearly one has to justify science and technology on a different basis than the fact that science is more advanced than, for example, sociology or the humanities. Scientists can no longer say, "We're going ahead, fellows, and we hope you can catch up." It seems clear that we cannot catch up. If we cannot catch up by stimulating the humanities and social sciences, maybe the best way to catch up is to slow science down.

Secondly, and more importantly, we are gradually diagnosing one peculiarly American disease which has as its primary symptom the belief not in E Pluribus Unum, but in the belief that we can get something for nothing. For two hundred and fifty years, America has been built on the belief that hard work, character, and industry somehow will give Americans the good life. And Americans believe that they do not have to pay for the good life or that the pay is the hard work and the industry. That belief has created habits that are very difficult to break.

The fallacy of this belief was recognized a long time ago, in relative terms, by a humanist who was perhaps ahead of the sciences. Goethe, at the end of the eighteenth century formulated and indeed built a number of rather interesting treatises around his concept of the principle of compensation. It is a humanist principle, and therefore it is difficult to weight it or analyze it or do anything but contemplate it. It simply and profoundly says that one does not get something for nothing. For every advance one makes, one pays. If we now begin to bring that principle forward and keep it constantly before our consciousness, we will then perhaps begin to close the gap between technology and the humanistic dialogues. If we can ever begin to think that for everything new that science, technology, and medicine gives us, we are going to pay, then I hope we will begin to ask, "What is the price?" The price sometimes may not be worth the advance. Then we will perhaps have to learn to live with a situation in which we give up our old habit of arguing that truth is a transcendent and absolute value, good in itself, and that we must never, never be prevented from reaching it. Perhaps now we may learn to live with the fact that being finite human beings, we cannot live with truth for very long without it destroying us.

In response to my paper, I have been asked to compare the Western approach to scientific education with other models, since I argued that among nations differences exist in teaching science. I have been asked whether the Russians, for example, do a better job than we do in integrating the teaching of science and philosophy.

I am by no means an expert on the difference between American and Russian scientific education, but a number of matters are worth consideration. First, America is almost consciously anti-philosophical. Philosophies do not build log cabins. America was built on pragmatism: if an idea works, it must be true, and one does not worry about ultimate philosophical consequences. That kind of pragmatism is a characteristic in American thought. One sees it; there are some aberrations, but the aberrations are interesting because they are aberrations—Emerson and transcendentalism yield rapidly to Pierce and pragmatism. America is a pragmatic nation, not a transcendental one.

The Soviet Union is not a pragmatic nation. Philosophy is of fundamental importance in Russia in that Marxist-Leninist-Stalinist thought lays claim at least to being science. It is a way of looking at the world and is of fundamental importance in the educational system. Russian youngsters in their earliest schooling are exposed to certain categories of philosophical thinking. Marx, in particular, is a very difficult philosopher. He drew his inspiration not only from Hegel but from a group of German philosophers at the beginning of the nineteenth century. Hegel himself was heavily influenced by this group, *die Naturphilosophen*. Their thought is difficult, intricate, and sometimes, I want to add, meaningless, but they were clever.

Russian youngsters, and particularly those who will study science, are introduced early in their lives to some philosophical problems, especially those concerned with the nature of space, the nature of time, and the nature of matter. These problems seem straightforward; everyone knows what space, time and matter are. Yet when they are examined closely in hard, specific terms, they seem to vanish. For a Russian student, the nature of space and time will be defined in Marxist-Leninist-Stalinist thought. If a youngster wants to be a good scientist, he should arrive at concepts of space, time, and matter that are compatible with Russian thought. Russia does not produce great philosophers, however, for there is an orthodoxy to which they must adhere. But at least Soviet students are made aware of philosophical concepts that I suspect are totally foreign to the average American undergraduate or graduate student. I know this is true, because I have taught physics majors in graduate school who minor in the history of science with me. When I ask them what they think of the Copenhagen Interpretation, they think I am speaking about Victor Borge playing Beethoven. They do not believe that we are discussing philosophy. Philosophy is in a sense an uncomfortable word for them.

I am not quite sure how to assess Soviet education, because when an educational system is locked into an orthodoxy, I do not think it is in a favorable position. I was however struck in one interesting device that the Russians use for reasons that will become obvious. In the sixth grade, or our equivalent, Russian youngsters begin to learn English by reading Michael Faraday's *A Chemical History of the Candle*. Using that book as a text is an incredibly clever thing to do. I do not know whether we teach Russian in the sixth grade, but we do teach English. We introduce our students to English by still having them read *Dick and Jane* or a modern variation on that story. I am still horrified that in the tenth grade the glory of English literature is—God help us—*Silas Marner*. The Russians are very smart. They would not give their children *Silas Marner* to read, but they give them instead *The Chemical History of the Candle*. Faraday uses simple prose for a kind of writing that is rapidly vanishing in America. Clearly written, the book raises scientific questions in terms of a candle, an everyday object in Russian households. The use of Faraday's book as a text for sixth grade is an example of a certain amount of imagination evident in Soviet education.

I would also like to compare modern, scientific education with an historical example. The example concerns the physical sciences in nineteenth century Germany. These sciences, especially chemistry until it declined in 1880, were considered philosophical sciences and were taught by the philosophical faculty. If a student wanted to study physics or chemistry before 1880, he went to the university to study them within a very consciously philosophical framework. He was exposed to the philosophy of Kant, Hegel, Schopenhauer, and Fichte, and his physics or chemistry was learned within that philosophical framework. It is no coincidence that Albert Einstein was trained within the German tradition and that he acknowledged his debt to the philosopher, Ernst Mach. Mach himself was merely the culmination of a long tradition suggesting that philosophical problems were to be found in classical mechanics. Germans began their questioning by asking what was meant by absolute space. The French did not go beyond a mathematical description of it, and the English claimed that everyone knew what absolute space was. The Germans, however, agonized over *Raum* and wondered what it was. As one begins to agonize over such a concept, it loses its clarity, and Newton's explanation of it does not look quite so satisfactory.

What is time? Again, the French could mathematize it, and the English knew it by common sense. Only the Germans realized that real problems arose when they asked that question. It is no coincidence that the educational system in Germany created the atmosphere which permitted even the young Einstein at sixteen to ask, "What would happen if I rode an electromagnetic wave?" Until one can ask a question, it is obviously difficult to find the answer to it. The questions we ask won't be determined by our education, but they will certainly be influenced by it. What will become a question that we can ask in public or what will be a question that we know our colleagues will laugh at will in large part be determined by the breadth of our education.

There have been some reforms in technical education, but they are very small, and they do not meet the real problem. I would never for a moment suggest that an educational system can create a genius. But how many geniuses started out in

Physics 101, and took the required sequence of Physics 101, 102, 110, 210, 211, 315, 316, and then said "To hell with it!" What is this nonsense? What is this dishonesty? All too frequently physics unfolds as a series of true and unquestionable statements about the world, supported by hundreds of homework problems, each with a right answer. Even the laboratory work is rigid and proscribed. Students all too often do not really have the opportunity to do any real experimentation until they are fifth year graduate students. Before that stage, they are forced to go into the laboratory to do arranged experiments which must yield the right, predetermined, answers. Unfortunately, this is what must be done today, for obvious reasons. A teacher cannot give an undergraduate the equipment necessary to do a meaningful experiment. He cannot tell the student that as a teacher he does not know the solution to a problem and allow the student to go to the lab to see what solution he can find. The student in Introductory Physics is assigned an experiment, for example, the measurement of the gravitational constant G. In this situation, more often than not, the student gets the handbook of chemistry and physics to look up the value of G to five decimal points. He then does his experiment, and he discovers that he does not get the correct value for G. So, he fudges it! The intellectual ability of the student is spent in figuring out a clever way to get the right answer.

I will use the field of organic chemistry to confess my own sins. The great stimulus of organic chemistry, when I took it, was to determine the identify and the quantity of impurities to add to chemically pure substances that we bought from a pharmaceutical house; the result was to be handed in as the laboratory product for the course. This task involved incredible amounts of work, and an analysis of the lab bench was only a part of the work. We had to know just how much lab bench to add to the product so that it would not look as if we bought it from a pharmaceutical house. I must confess, that I have begun to wonder if my instructors were not diabolically clever. Did they not know that that was exactly what we were going to do, and was that not the way the task had really been set up? I learned an amazing amount of organic and inorganic chemistry that way, because the lab bench was inorganic. However, to call that work an experiment, to call it anything but cooking, seems to me to be dishonest.

We may have lost some first-rate minds because of this system of education. If I had had a first-rate mind, I think it would have been turned off by the end of my sophomore year, and maybe I would have done something else. There is no requirement that one continue with a course of study. One should be free to change fields to do what one wants. One of the lovely things about Einstein's education was that he did, in fact, do what he wanted. He wandered through various schools, and he studied what he wanted and paid no attention to anybody requiring him to do problem sets or lab bench analyses. He gathered knowledge together by himself, and when he needed help, he sought it out. If Einstein had been subjected to the American educational system, he would never have been allowed to pursue the independent study he did in Europe.

A part of the American dream which has turned into a nightmare is the idea that education can do everything. I do not want to suggest that idea at all. I merely want to suggest that one of the things that we should not want education to do is to stifle the few talents we have. As far as I can see, we are not doing a very good job of making sure that these talents are not being stifled. We have developed a complicated science and technology with an emphasis on problem-solving that has grown in importance and now dominates our lives. The expression that ninety-eight percent of all the scientists who have ever lived are living today is, I am sure, familiar. The expression is probably true, unless one wants to accept my analysis of science which suggests that what keeps science alive is a relatively small number of people, probably constant in any given age. What is dominant today, on the other hand, is the problem-solver who sees nothing but problem-solving and who therefore insists that that is all that can be seen. And then there are those few people who are just turned off by problem-solving, and perhaps just one or two of them may be a Faraday or an Einstein.

BIBLIOGRAPHY

Bernstein, Jeremy, *Einstein* (The Viking Press, M21, 1973).

Clark, Ronald W., *Einstein, The Life and Times* (Avon Books, M117, 1971).

Koestler, Arthur, *The Sleepwalkers* (see especially Part Four on Kepler, *The Watershed* (Grossett's Universal Library, UL 159, 1959).

Kuhn, Thomas, *The Copernician Revolution* (Harvard Paperback, HP16, 1957).

Kuhn, Thomas, *The Structure of Scientific Revolutions*, 2nd ed. (University of Chicago Press, IEUS II/2, 1970).

Schilpp, Paul Arthur (ed.), *Albert Einstein, Philosopher-Scientist*, Vol. I and II (see especially Vol. I, Einstein's *Autobiography*) (Harper Torchbooks, TB502 and TB503, 1951).

Williams, L. Pearce, *Michael Faraday, A Biography* (Clarion Book, 1971).

ARE
THERE
TWO
CULTURES?

INTRODUCTION

Although the first section of this book stresses the distinctions between science and technology and the implications that these distinctions convey for the future of science, technology, and society, the growing gap which exists between the current scientific-technological community and the rest of society was only mentioned in passing. This gap between the "scientific culture" and the "literary culture" was emphasized over ten years ago by Lord Snow in his now famous Rede Lecture at Cambridge on "The Two Cultures." The ensuing years produced a tremendous volume of commentary and discussion of Lord Snow's assertions. He responded first with surprise, then chagrin, over the uninformed and careless commentaries by many of his critics and finally formalized his response in *The Two Cultures: A Second Look*, in 1963. In this essay, Lord Snow maintained his original contention that advanced Western society had "lost even the pretense of a common culture." He offered a précis of his earlier position as follows:

Persons educated with the greatest intensity we
know can no longer communicate with each other
on the plane of their major intellectual concern.
This is serious for our creative, intellectual, and
above all, our normal life. It is leading us to
interpret the past wrongly, to misjudge the present,
and to deny our hopes of the future. It is making it
difficult or impossible for us to take good action.

Lord Snow characterized the two cultures as "the scientists,
whose weight, achievement, and influence did not need
stressing," and "the literary intellectuals," who

represent, vocalize, and to some extent shape and
predict the mood of the non-scientific culture:
they do not make the decisions, but their words
seep into the minds of those who do. Between
these two groups—the scientists and the literary
intellectuals—there is little communication and,
instead of fellow-feeling, something like hostility.

Lord Snow was convinced that he was offering a generalized
description of a state of affairs in modern Western society
which he deeply disliked, but which seemed not only to
exist, but to be a persistent feature of our society.

The second section of this book is a discussion of and
confrontation with Lord Snow's theme. "Are There Two
Cultures?" is meant to re-examine the contention of the
existence of the two cultures a decade after Lord Snow's
essay was written. It develops the possibility of a line of
argument which pursues the implications of the separation
within the "scientific" community outlined in the first
section of the book. George V. Cook begins by distinguishing
between the "scientific" and the "literary-intellectual" communities as Lord Snow described them.

In examining the validity of Lord Snow's contentions, Mr.
Cook states that accepting the dichotomy between the

scientific and literary-intellectual communities produces a false view of our society. Convinced that our Western society consists of many different cultures, Mr. Cook nevertheless suggests that the dialectic between Lord Snow's two cultures presents an interesting and perhaps useful intellectual framework from which to address the problems of technology and its relationship to our society.

Mr. Cook is less concerned about the existence of different cultures than he is about the degree of polarization that exists among them. The differences are not necessarily destructive. Instead, the resulting tensions are what "makes our society grow and it is central to the evolution of civilization." Although one may regret the degree of separatism and polarization, one need not accept "the premise that a single homogeneous culture of intellectuals and scientists would be a preferable state."

Mr. Cook views the division between opposing cultures as presenting the framework for the tension that brings a dynamic quality to civilization. If that division is too deep, as it may be in the United States, it is potentially dangerous for humanity. "While a civilization evolves from tensions, it may be destroyed by polarization." Although the response to the tensions may eventually be judged to have provided a beneficial impetus to civilization, the present unresolved divisions and tensions represent serious, immediate problems. Borrowing from his co-worker, Henry Boettinger, the phrase— "greater interaction and stronger couplings"—Mr. Cook suggests that if properly managed, there could be increased and systematic communication between the various segments of society with a resulting interdisciplinary exposure.

Like L. Pearce Williams, George Cook calls for a reform of our educational system. Although Professor Williams envisions an educational system that allows man's best creative efforts not only to break through but also to be energetically developed, Mr. Cook views education in part as a methodology for developing increasingly effective, problem-solving scientists—imbued with a vision that is not necessarily religious but certainly is humanistic and humble—who are able to use a systems approach with interdisciplinary insight. He stresses that while scientists have the right to specialize, they also have the duty to attempt to understand and communicate with people in other disciplines. The resolution of the polarizations arising from the separatism induced by a multiplicity of cultures will, he feels, determine the fate of Western society.

Mr. Cook, unlike Professor Williams, places great emphasis upon the differences between the creative man, whose energies are the product of inner drives and of a broad, total view of the world, and the problem-solving man. This difference is different in emphasis from the distinction between the scientific and literary-intellectual communities stressed by Lord Snow. As Raul Hilberg suggests in his commentary on Mr. Cook's paper, there is an enormous similarity between the creative efforts of the scientists and of the literary man. Both the poet and the scientist view the world from a cosmic perspective and are not particularly interested in manipulating parts of it without reference to the whole.

Mr. Cook concludes his appeal for decompartmentalization and interdisciplinary involvement by reminding us that no one is omniscient, and he asks rhetorically, "Don't we all need an interdisciplinary sense of participation and relevance if we are, as individuals, to attain our maximum personal satisfaction and contribution to the needs of our society? Don't we all need a vision?"

The nature of this vision provokes much of the commentary on Mr. Cook's paper. Cornelius L. Coyne, Jr. agrees with the basic thrust of Mr. Cook's paper concerning the necessity of meeting the most pressing problems of modern society by better organization of human and technical resources. Accepting Lord Snow's thesis, Mr. Coyne refines the suggestion of the existence of a third culture which is comprised of the social sciences. The real problem of this third culture, according to Mr. Coyne, lies in its inability to establish priority and develop meaningful applications of the existing technology.

The question of priorities concerns Jeremy P. Felt. He asks, "Who is to decide when 'tensions' or 'cultures' have become 'polarized' and *by what standards* the merits or demerits of each are to be judged?" In raising the question of who establishes the priorities by which the technology of Mr. Coyne or Mr. Cook will be applied, Professor Felt voices a fundamental issue. Although Professor Felt offers no answers to his own question, other contributors suggest that the answer may lie in a commitment to a metaphysical construct, which might help guide the application of technology. None of the authors, however, directly confront Professor Felt's question, for even the adoption of a metaphysical position presupposes a process of selection.

Professor Hilberg's commentary makes this dilemma starkly clear, with its vast portent and strong hint that it may, indeed, be too late to reverse the direction of civilization. Creativity, the very stuff of science, according to Professor Williams, is also the same kind of creativity of Lord Snow's literary-intellectual culture. If, as Professor Hilberg suggests, the literary culture in contemporary society has been submerged and that writers like scientists are unintelligible to the public, it then may be too late to wait for the development of a new metaphysics. It may even be too late to bring the necessary interdisciplinary fertilization to the teams of problem-solvers in whom Mr. Cook places his hope and confidence.

ARE
THERE
TWO
CULTURES?

George V. Cook
Vice-President
Western Electric Company

I have a certain ambivalence, as a lawyer, about contributing a paper to a book concerned with the role of science in our modern culture. My early education, however, was in science and mathematics, and I believe that it is not without significance that I learned my calculus from an illustrious American composer of music, Milton Byron Babbitt. That a young man learned his calculus from a musician would have delighted C.P. Snow, especially in view of the cultural divisions and educational problems which I want to discuss in this paper.

In his now famous Rede lecture(1) on the two cultures, Lord Snow argued that modern intellectual society has become polarized into two mutually exclusive intellectual communities: the "literary" and the "scientific" cultures. In the literary culture, Snow groups novelists, poets, playwrights, and especially literary scholars. The scientific culture is epitomized by physical scientists.

The "literary intellectuals," particulary the existentialists, obstinately refuse to concern themselves with contemporary scientific development. As a result, they fail to perceive the relevance of scientific developments to the problems still afflicting most of mankind. The literary intellectuals thus tend to be increasingly "anti-scientific," and they have never, according to Snow, "tried, wanted, or been able to understand the industrial revolution, much less accept it."

On the other hand, members of the growing "scientific" culture facilely excuse a total ignorance of the traditional literary culture. To them that culture is irrelevant to their problems and moribund to boot. It is not surprising that the "literarists" regard the "scientists" as "socially impoverished."

Each group, according to Lord Snow, has become "tone-deaf," and each increasingly accuses the other of being "anti-intellectual" and oblivious to their fellow man's condition. He sees a widening "gulf of mutual incomprehension," sometimes "hostility and dislike, but most of all lack of understanding." In addition, he finds the separation "much less bridgeable among the young than it was even thirty years ago." He finds this growing separation destructive.

In the rest of his essay, Lord Snow orchestrates his theme with a good deal of counterpoint, some dissonance, and concludes with a coda devoted to the need of improving communications and, especially, the content of our educational system in order to break down the invidious trend toward "parochial cosiness." It is an ironical coda, for Lord Snow's own career bridges both cultures. As he has said, "By training I was a scientist; by vocation I was a writer."

I face the uneasy task of not appearing to disagree very strongly with Lord Snow, for there is much wisdom in what he says. The division that he finds is not universal, yet it does exist, as the following colloquy between Robert Oppenheimer and Paul Dirac, described by N.P. Davis in *Lawrence and Oppenheimer*, indicates:

"Why do you waste time on such trash?" Paul Dirac asked (Oppenheimer) irritably a month after he started spending two hours a day studying Dante. Dirac had a great hard bare mind then wholly dedicated to thinking out the mathematics of a negative energy-state which nobody before him had conceived of. Dostoevsky, Proust, and Thomas Acquinas were other indulgences Dirac disliked to see in a colleague who had proved capable of actually understanding and contributing to the Dirac Theory. "And I think you're giving too much time to music and that painting collection of yours," Dirac would say during long walks he took with Oppenheimer for the purpose of persuading him to quit the pursuit of the irrational.

I do, however, have difficulty in accepting the sharpness of the division that Lord Snow finds between the two cultures. Lord Snow himself was a scientist as well as a novelist, and Robert Oppenheimer was a poet as well as a nuclear physicist. My argument with Lord Snow is not over the existence of two cultures, but over the schism he finds, taken in the broader context of history and of our contemporary society.

First, it must be recognized that Lord Snow was commenting primarily on the English scene and that he observed in *Second Look* that in the United States "the divide is nothing like so unbridgeable." Second, it should be recognized that Lord Snow's choice of two cultures was itself a dialectic choice, i.e., one made for the purpose of argument. His choice was candid if oversimplified, provided useful counterpoint, but necessarily precluded some deeper insights. Lord Snow in a sense recognized his choice of two cultures for what it was in *Second Look*, and he dismissed it by saying any further refinements would be "subtilizing." He stated:

Now for the number two. Whether this was the best choice, I am much less certain. Right from the start I introduced some qualifying doubts. I will repeat what I said, near the beginning of the lecture. "The number two is a very dangerous number: that is why the dialectic is a dangerous process. Attempts to divide anything into two ought to be regarded with much suspicion. I have thought a long time about going in for further refinements, but in the end I have decided against. I was searching for something a little more than a dashing metaphor, a good deal less than a cultural map, and for those purposes the two cultures is about right, and subtilizing any more would bring more disadvantages than it's worth."

Significantly and pertinently, Lord Snow felt, on further reflection, that a different division might have been drawn between "pure science" and "technology." Certainly, the division between some of today's scientists and technocrats is equally deserving of our attention. For example, Bell Telephone Laboratories, as indicated in H.W. Bode's study, has provided an unusually interesting institutional catalyst for coupling these disciplines, and as Sir John Cockcroft found, provides a most useful locus for further study of the factors that serve to provide a common interest among the "pure" scientists, the "applied" scientists, and the "technocrats."(2)

Lord Snow's assertion that it is disadvantageous, analytically speaking, to consider a plurality of cultures is, as he admits "in a sense . . . true; but it is also meaningless."(3) I cannot be as simplistic and unequivocal. I recognize Lord Snow's two cultures, but what seems to me important is that there are not just two cultures; there is an infinity of cultures in our society.

To ignore the existence of a plurality of cultures or to say that they are "meaningless" is to obscure some very fundamental considerations relating to the significance of Lord Snow's basic thesis.

One can categorize our society in infinite ways: literary

intellectuals and scientific intellectuals; scientists and technocrats; lawyers and engineers; poker players and chess players; Don Quixotes and Sancho Panzas; the culture of the so-called "Establishment" and the counter-culture of Charles Reich. The existence of these diverse cultures and the tensions they generate are what makes our society grow and are central to the evolution of civilization. One can thus fully sympathize with Lord Snow and lament the degree of separatism and polarization that he found, without accepting the premise that a single homogeneous culture of intellectuals and scientists would be a preferable state.

Civilization is the product of tensions. Specialization requires some degree of separation, and separation implies a degree of tensions among specialties. The degree of homogeneity or the synthesis that results ultimately from these tensions is an integral part of the process of civilization. To fail to recognize the importance of tensions is to miss a central point of the history of our civilization or of any civilization. It follows then that the existence of scientific and literary cultures is no more destructive than the existence of any other two cultures.

Jean Francois Revel makes this very point in his commentary on contemporary America:

> America in 1971 is a mobile entity. It crosses all lines, not only financial, social, and familial lines, but also cultural and moral lines. And, containing as it does a diversity of cultures, and contradictory moral systems, it generates collective and individual crises with constantly increasing frequency. *It is precisely these crises—which are numerous, permanent, and always new—that comprise a modern revolution*; that is, a revolution as it is realized in societies that are too complex, and insufficiently hierarchical and centralized, to be changed overnight, by a single coup and in a single direction. Crisis has become America's second

nature. But, in order to realize this fact, one must live in it; and that is why the rest of the world perceives only dimly the true dynamism of present day America. (Italics added.) (4)

I partially agree with Revel's observations, but I also share Lord Snow's basic concern that the gulf between scientific and literary cultures, even here in the United States, has grown too wide. It has grown so wide that it contributes unnecessarily to the malaise of polarization overhanging our society. What concerns me is not the existence of different cultures but the degree of polarization that does exist. While civilization evolves from tensions, it may be destroyed by polarization.

The focus of inquiry should be the ways in which we can facilitate the resolution of the tensions inherent in our different cultures. Is it not the way in which we resolve these tensions or fail to resolve them that will provide the criteria by which historians and anthropologists will judge our civilization?

The events of the past decade have confirmed the growing gulf of mutual ignorance which Lord Snow discerned and regretted. However, these events have also shown that rigid differences over time tend to be blurred and that different groupings take their place at polar extremities. Some regrouping, therefore, seems to be in order.

One does not need to be a literary scholar or a scientist to perceive that the gulf between many scientists and non-scientists continues to be a matter of grave concern and raises all sorts of paradoxes. Environmental pollution, DDT, phosphates, hexochlorophene, the goal of a "zero-growth" society have all been popular topics of contemporary "intellectual" criticism. Scientists and engineers have ignored for too long the significance of "waste" in their cost equations, while many literarists continue to heap scorn on technology in a modern form of "know-nothingism" without perceiving technology's message of hope.

Peter Drucker, a noted member of the "third culture," amply confirmed Lord Snow's thesis and the inherent dangers of the difference between the two cultures in the January 1972 issue of *Harper's* in which he pointed out:

> I happen to be a member in good standing of the Sierra Club, and I share its concern for the environment. But the Sierra Club's opposition to any new power plant today—and the opposition of other groups to new power plants in other parts of the country (e.g., New York City)—has, in the first place, ensured that other ecological tasks cannot be done effectively for the next five or ten years. Secondly, it has made certain that the internal-combustion engine is going to remain our mainstay in transportation for a long time to come. An electrical automobile or electrified mass transportation—the only feasible alternatives—would require an even more rapid increase in electrical power than any now projected. And thirdly it may well, a few years hence, cause power shortages along the Atlantic Coast, which would mean unheated homes in winter, as well as widespread industrial shutdowns and unemployment. This would almost certainly start a "backlash" against the whole environmental crusade.(5)

The same persistent paradoxes can be found with respect to the hostility, reflected in a large segment of the intellectual commentaries, toward insecticides. Rachel Carson could have had no conception of what she wrought, as Peter Drucker states:

> Today, for example, no safe pesticides exist, nor are any in sight. We may ban DDT, but all the substitutes so far developed have highly undesirable properties. Yet if we try to do without

pesticides altogether, we shall invite massive hazards of disease and starvation the world over. In Ceylon, where malaria was once endemic, it was almost wiped out by large-scale use of DDT; but in only a few years since spraying was halted, the country has suffered an almost explosive resurgence of the disease. In other tropical countries, warns the U.N. Food and Agricultural Organization, children are threatened with famine, because of insect and blight damage to crops resulting from restrictions on spraying.(6)

Our problems are far from easy; they involve challenges of such a magnitude that we need to enlist all of our resources and talents to find solutions for them. They simply cannot be left to the individual domains of "the intellectuals," "the scientists," or "the contemporary populists."

We can however thank the environmentalists for confronting us with environmental risks and problems. Our awareness is a start; but it is simply not enough. We need foremost to bridge the gulf of mutual ignorance and pride. We need Alvin Toffler's recognition that the old glue which proved adequate for a simpler society is no longer able to withstand new and stronger forces of "future shock" that have ruptured old relationships among individuals, organizations, and cultures.

Different cultures or disciplines can either cause friction, abrasion, or impedance when brought into contact, or they can be used to link the strengths of one to offset the weaknesses of the other. Although not put so tersely, this is the real message that is ultimately conveyed by Lord Snow, and I concur with it.(7)

My associate, Henry Boettinger, and I like to refer to "the politics of creative tension." Boettinger applied this concept to the problems of contemporary business management(8), but it is equally applicable to broader aspects of our society. In another context, Lord Snow speaks similarly of the need to facilitate "creative chance."

Lord Snow, of course, was not the first to discern or lament the growing division between the arts and sciences. Several centuries ago, Sir Francis Bacon, another lawyer, opined with respect to his *prima philosophia*: "I hold it for a great impediment towards the advancement and further invention of knowledge, that particular arts and sciences have been disincorporated from general knowledge . . . "

Many historians date the emergence of the "scientific method" from Bacon, as well as the disciplines which mark the beginning of the modern scientific culture. Yet the evolution of that culture was itself the product of its own Hegelian synthesis.

Bacon lived in an age when it was possible to reside comfortably in the two cultures. For us it scarcely possible in this age of specialization to even keep apprised of the advance sheets and periodicals of our own callings, not to mention those of other disciplines. I believe it was Lord Snow who said that "modern engineers are terribly inclined to accept the culture in which they were born." That comment can be applied to any specialist.

If we are to have the many benefits of specialization, there will always be a gulf between specialists. The greater the degree of specialization, the lower will be the degree of receptivity to information from beyond the circle of the specialist's own discipline.

What we need then are new understandings and new glues to hold us together; we need greater interactions and stronger couplings. This need should not be difficult for the scientist to perceive. Everything Lord Snow said is an illustration of the Second Law of Thermodynamics: every closed system tends to run down and become less and less organized as time goes on. Physicists say, "Entropy increases." They also say that for the tendency to be reversed, energy must be brought to the system in a certain way.

Although the great and whimsical James Clerk Maxwell overcame the Second Law with his "demon," the question for us is, "Where shall *we* find the source of the needed energy?" It cannot be generated from any one culture or any one discipline; it must be a synergistic phenomenon. We need the "hot" creative ideas of all cultures.

We need a far greater application of the talents and "hot" creative ideas of the scientists and engineers than has been evidenced in the past. I come from a business in which the concepts of "systems engineering" are so deeply implanted that they are second nature. The scientists and the engineers are the foremost repositories of this discipline which sorely needs a wider interdisciplinary exposure.

The late Jack A. Morton of Bell Laboratories, more than any intellectual or scientist I have known, ceaselessly carried the message throughout the world that the innovation process is not just a technical or technological process. It is the process of renewal; it is above all a people-process; and its disciplines sorely need to be adapted to social imperatives as well.

We need to cast off what John W. Gardner called, "the excessively narrow technological conception of innovation," (9) and to convert the innovation process into a powerful technique for humanizing the future. As Glen McDaniel of Litton Industries recently said, "To my mind the most exciting aspect of the systems concept is this striving to accomplish something never before possible." (10)

The "systems concept and method" is not understood by the intellectual community at large and by a substantial sector of the scientific community as well. However, it is beginning to be applied in city management, in housing, in the management of public education, and in other socially active areas. Its proponents include the "Think Tanks" at the Rand Corporation, at Stanford Research Institute, and at The Institute for the Future. It is indeed a hope for the future; its techniques must be expanded from a single disciplinary approach to a multi-disciplinary approach, from technical matters to social imperatives. We have begun in a tentative and exploratory way, but at least we have begun.

I want to turn now to my concern with the content of

education. I grew up in an academic environment in which every engineering student was required to participate for two years in a liberal arts education. That liberal education for a perspective engineer certainly facilitated understanding and communication later. It does seem to me that we tend to specialize much too early in our education. To paraphrase John Updike: "We all contain chords that someone else must strike." (11)

Anyone who knows me recognizes that I not only preach interdisciplinary attitudes, but that I am proudest of my own organization because of its interdisciplinary approach to problems. A department called "Regulatory Matters" may sound pedestrian, but it is not. "Regulatory Matters" deals not only with internal business matters but also with external problems of our organization which require a staff of economists, lawyers, engineers, accountants, and mathematicians. We had found that traditional methods and a traditional staff were inadequate to solving many business problems.

At first, creating a staff with various interests and specialties raised havoc with some members of our legal and operating departments. The new staff should have "shaken up" our older departments, but we have shown that it functions effectively for the organization. More importantly, we have provided management with insights that never previously emerged from the traditional staff organization. Best of all, this interdisciplinary approach has created an environment of participation and made otherwise "compartmentalized tasks" challenging and interesting.

I once had a professor of "evidence" who taught the legal intricacies of this technical branch of the law as a course in logic, with the use of Greek symbols. He gloried in referring to his course in the syllabus as a branch of "Epistemology," or as that part of philosophy which dealt with the "theory of knowledge." I could never have passed the New York Bar Examination on the basis of what he taught me, but I am a better man for the experience in his classroom. I will always remember that we were never permitted to refer to "a fact" as "a fact," but could only say what we perceived was a "proposition about a matter of fact." The fact was an "ultimate probandum" or the thing to be proved. What we discerned was a "probans," an intermediary step in the determination of the truth.

The point is simply that no one has a monopoly of "knowledge." Each of us in our own way can advance the human accumulation of knowledge as we perceive it, but no one is omniscient. The truth has many facets and, as in a diamond, not all facets may be seen at once or from a single angle. We all learn from each other; we are all building Bacon's pyramid of knowledge.

Finally, I want to make some observations which are in some measure the legacy of my liberal arts education. In the Socratic tradition, I have chosen to intersperse some questions throughout my text.

I do think that we all need the perspective provided by the literary scholars of Lord Snow's two cultures, if for no other reason than to realize that other less advanced societies accomplished herculean feats in their day. It occurs to me that we sometimes tend to be overly awed by contemporary science and technology. Given the resources of the time, one can only ask, which of the following was more difficult to accomplish: the building of the pyramids or Chartres, or the landing of a man on the moon?

If that question disturbs the scientists and engineers a bit, I would also ask whether our literary colleagues do not need to recognize that the only real hope for the poor rests ultimately in modern science and technology? We can all criticize the threatening aspects of technology, but isn't the only weapon we have to oppose the bad effects of technology, technology itself?

Don't we all need the awareness, taught so well by the Greeks, that no one is omniscient, not even a Newton, a Beethoven or an Einstein? Don't we all, including our literary scholars, need the sense of humility that is taught in Samuel Eliot Morrison's great epic, *The European Discovery of*

America? Don't we all need an interdisciplinary sense of participation and relevance if we are, as individuals, to attain our maximum personal satisfaction and contribution to the needs of our society?

Don't we all need a vision? If not the vision of Moses when he saw God on Sinai, then don't we need our own individual, pragmatic visions, like Jeremy Bentham's utilitarian vision of the "greatest good for the greatest number?" Finally, don't we all, in this time and place, need the optimism not of Candide, but of a Revel, a Merton, or a Van Doren?

We each need to cultivate our own garden, but we also need to recognize, in the terms of Hohfeldian legal analysis, that for every right there is a correlative duty. And while we have the right to specialize, we also have the duty to attempt to understand and communicate with others. I am grateful to Lord Snow for reminding us of that right and that duty.

NOTES

(1) The "The Two Cultures and the Scientific Revolution" was a lecture delivered at Cambridge in 1959 and published by the Cambridge University Press. "A Second Look" was first printed in 1963 and included the second edition of "The Two Cultures" under the title *The Two Cultures and the Scientific Revolution: and a Second Look*, published by Cambridge in 1964. Fortunately, these commentaries, together with several other important statements of Lord Snow, have recently been republished by Scribners under the title, *C. P. Snow—Public Affairs*.

(2) *The Organization of Research Establishments*, Cambridge University Press, 1965. See also Prof. H. W. Bode, *Synergy: Technical Integration and Technological Innovation in the Bell System*, Bell Laboratories, 1971.

(3) Second Edition, p. 66.

(4) *Without Marx or Jesus*, Author's Note in reply to Mary McCarthy's Afterword, p. 262, Doubleday, 1971. Compare in this connection, Toffler's thesis in *Future Shock*,

Random House, 1970, with respect to the imperatives of accelerating change, and Galbraith's statement: "The imperatives of technology and organization, not the images of ideology, are what determine the shape of economic society." *The New Industrial State*, Houghton Mifflin, 1967, at p. 7.

(5) *Saving the Crusade*, Harpers, January 1972, at p. 67. For a similar commentary as to the impact of a "no-growth" society on minorities, the Third World and the environment, see *A World Without Growth?*, Henry C. Wallich, N. Y. Times, Feb. 12, 1972 at p. 29, which echoes Walter Heller's earlier observation that we must grow awfully fast "just to stand still."

(6) *Ibid.*, pp. 70-71. See also Prof. T. H. Jukes, *Like It or Not, DDT is Good for You*, N.Y. Times, Aug. 3, 1971, at p. 29.

(7) See *A Second Look*, Second edition, at p. 98.

(8) *A Design for Business Vitality*, the British Institute of Management, September 1971.

(9) *Self-Renewal*, Harper & Row, 1963, at p. 30.

(10) *The Meaning of the System Movement to the Acceleration and Direction of the American Economy*, Mimeo, 1968, at p. 4.

(11) *Rabbitt Redux*, p. 72.

COMMENTARY

Jeremy P. Felt
Chairman, Department of History
University of Vermont

I am not a scientist but a historian interested in some of the patterns of social and intellectual change in the United States. I would also like to say that while I am going to disagree strongly with some of the statements and assumptions in Mr. Cook's paper, I think he deserves thanks for raising so forcefully some of the ideas which are important to one key segment of our society.

Although he states that "there is an infinity of cultures in our society" and that these diverse cultures are what makes our society grow, Mr. Cook really favors only diversity of a rather restricted kind. For example, while he recognizes that specialization requires some degree of separation and that these separations act to set up creative tensions, he finds it essential that these tensions stop short of becoming polarized. He urges that we facilitate the resolution of the tensions inherent in our different cultures and, indeed, suggests that our success or failure in doing this will be the standard by which future historians judge us. The unanswered questions are, of course, *who* is to decide when tensions or cultures have become polarized and *by what standards* the merits or demerits of each are to be judged. Otherwise it would be a simple matter for the controllers of power in society to create approved or "co-opted" dissenting groups who, because *they* did not question the central premises of the group in power, were engaging in creative tension, while groups that raised fundamental questions, or rejected the premises out of hand, would be accused of "polarizing" the situation. It seems to me that Mr. Cook's position could easily lead to the principle of "disagree but stay on the reservation which *we* have defined." After all, when we are all having creative tension by debating the best way to invade France, who wants someone polarizing us by asking whether we should invade France?

Up to a point in the paper it was not entirely clear who was to perform this kind of decision-making. For example, at one place the author urges that different cultures or disciplines be used to link the strengths of one to offset the weakness of the other. This seems to me to be in contradiction to C. P. Snow's statement that each side of his cultural dichotomy fails to understand the strengths and weaknesses of the other. How then can they (or any multiplicity of cultures) link their strengths and weaknesses? I don't know the answer, but I deduce from material toward the end of his paper that Mr. Cook envisions a certain kind of person achieving this linking. He is the technocrat, the expert, the person who believes, in Theodore Roszak's words, "that whenever social friction appears in the technocracy it must be due to what is called a 'breakdown in communication' . . . thus we need only sit down and reason together and all will be well." Mr. Cook's conviction here is clear, at least to me. It shows up in his admiration for the systems concept and for the "think tanks" and in his clear commitment to the expansion of the techniques of those questionable enterprises from technical matters to social imperatives. It is present also in his statement about encouraging "hot" creative ideas as opposed to "cold" ones. What is a cold creative idea and who, by what means, would discourage it?

Moreover, Mr. Cook is clearly quite optimistic about the scientific practitioners of systems concepts being able to lead us into a grand new future. He may even believe, though I would expect him to rebut this, that science can solve all our problems if we just give the technocrats free rein. I was somewhat disturbed by his relegation of the perspective provided by literary scholars, including, I assume, historians, particularly non-computerized ones, to the role of reciting the glories of the past, at least the technical glories of the past. I'm not sure where that leaves the workman on that magnificent medieval cathedral who created that piece of stained glass so high that no one need see it because God could see it. That workman might be a problem in a modern automobile factory. I was even more disturbed by Mr. Cook's statement that the only weapon we have to oppose the bad effects of technology is more technology. I am reminded of the famous sorcerer's apprentice scene in Walt Disney's "Fantasia," where a similar procedure on the part of the sorcerer led to a fatal flood.

At the end of his paper Mr. Cook suggests that if we don't have Moses' vision of God on Sinai—where, incidentally, God took a most irrational and unsystematic form in speaking from within a burning bush and the perception of God's message required that Moses *suspend* his rational faculties— we could at least have a Benthamite vision of the greatest good to the greatest number, said vision to be achieved, presumably, through continued economic growth. It has now become fashionable to argue that those who oppose growth are either unaware of poverty in the world or that they are elitists saying "let them eat cake" and secretly opposing letting anyone else enjoy their high standard of material living. The argument that we can only end poverty through more growth and technology is a powerful one, and it will undoubtedly be one of the major issues of the next decade. I know only that we *now* possess the means to end poverty, certainly in this country, through wealth redistribution on even a relatively small scale. I am not confident that either our environmental limits or our human limits can stand an increasing reliance on manipulative technique. If we do not soon learn that the idea of objective consciousness is as powerful a myth as that of Osiris, Ra, or anything else, we are in for a very rude awakening. Objective, rational, scientific consciousness is, I submit, *one* way of looking at reality. It is not necessarily the best way to look at all reality nor is its current prevalence necessarily evidence of progress. As Roszak persuasively argues, the Nazi enterprise was populated not by Wagnerian hero types but by rather banal and petty dull technocrats.

In summary, I am in favor of an infinity of cultures—even assuming that with a finite number of people such an infinity is possible. I am not however in favor of any social approach which overtly or covertly assumes that such a multiplicity can be benevolently overseen by a regime of supposedly value-free experts whose sole aim in life is purportedly to promote a happy and creative diverse world. If this were the seventeenth century and I were a Puritan, I would probably say that such an approach was a revolt against God and an effort to set up a planning board in his place. Since this is the last half of the twentieth century, I am terribly conscious of what we *do not know* about human beings and very pessimistic about the possibility of systems analysis even leading us to the right questions.

COMMENTARY

Cornelius L. Coyne, Jr.
General Manager, Information Systems
Western Electric Company

In trying to view the polarity or gulf between the scientific and literary cultures, I find myself somewhere in this infinity of cultures between the two poles. Lord Snow labeled this place "the third culture." Between the two extremes, I find that in certain respects there is a great sameness within these two diverse groups, particularly in the techniques that they employ to practice their professions. The novelist for example uses his powers of observation for his craft and it is imperative that he observe the patterns of society. He classifies and catalogues his observations to compare them to some standard. Is this process any different than the one employed by those scientists who observe nature and then attempt to classify and catalogue their observations? Each group clearly anticipates that whatever observations they make will contribute to the body of knowledge and will thus find some useful place in society. In many cases it is the third culture that is the recipient of the product of the other two cultures.

Mr. Cook suggests that the only real hope for the poor

rests in modern science and technology. I agree that the best hope lies in the application of technology. This is not a simplistic statement. Much of the application of technology lies in the hands of those who reside in the third culture, and it is our use of the fruits of the scientific community that the literary fraternity so often challenges. Despite often undeserved and frequent criticisms, technical solutions exist for such problems as food supply, mass transit, and water and air pollution. The real problem lies in the inability of the third culture to establish priorities and develop meaningful applications of the existing technology. I want to suggest from the vantage point of the third culture that the separation between the scientific and literary cultures is the result of the inability of the third culture to use constructively the ideological, human and material resources at hand.

As long as we fail to resolve these difficulties, we shall continue to exist in a state of tension which contributes to the creative process. But if this tension exceeds the yield strength of the society, the society will be torn apart. We must have the strength to be able accurately to observe all events and then establish a balance between the opposing forces. To do so will not be easy, and in fact the very process of observation may be a disturbing action not unlike the Heisenberg statement of uncertainty as it applies to the world of subatomic particles. In conclusion, Mr. Cook's closing observations or challenges are properly directed as much to the third culture as to the first two.

COMMENTARY

Raul Hilberg
Chairman, Department of Political Science
University of Vermont

C. P. Snow's lecture, *The Two Cultures*, was delivered in 1959 and was later published in fifty pages, thirty lines to the page. It is quite a bit longer than the average contribution to a learned journal, even a social science or humanistic periodical. It is a good deal shorter however than what publishers expect in terms of length for a book. The lecture falls into that category of literature which is comprised of manifestos, lectures, or essays that are, in Lord Snow's own words, "ripe for their time," because they state a central proposition. Secondly, the title, *The Two Cultures*, with its short, majestic words, expresses the author's entire thesis. How very seldom does one find titles like that in literature. It is no wonder that thousands of miles from the place where the lecture was originally given, we are still considering it years later, amending it, modifying it, adding to it, and at the same time affirming its significance.

I will praise Lord Snow without any ambivalence, without any reservations about him, but with great ambivalence and with many reservations about the society he described. I have no trouble with the two extremes that he poses, because it is quite easy to imagine two polarities—the literary culture and the scientific one—and between them a host of disciplines. These disciplines may be called a "third culture," but the important thing is that they are between the two poles. They may be technological arts, like photography, film, architecture, all of which utilize knowledge or the physical phenomena that produces artistic effects. On the other hand, they may be behavioral sciences which seek to reduce man to a system and society to a social machine. At both poles we are left with the solitary phenomenon of the novelist, the poet, or the physical scientist. There are two worlds: one verbal, one mathematical; one literary, one scientific. They are central from a structural point of view. They ring true, and they do so most especially because Lord Snow was a practioner in both of these extreme areas.

It is significant that Lord Snow's greatest detractors, the British critics, F. R. Leavis and Michael Yudkin, chose for their title almost the same words as Mr. Cook did for his paper. Leavis and Yudkin did not deny the validity of anything Lord Snow said; they questioned only Lord Snow's credentials. Lord Snow, they claimed, was not a great man in both literature and science. Professor Williams refers to the solitude of a scientist, the rare man with a cosmic vision. Poets and novelists also labor with that kind of solitary vision, viewing the world in its past and future. Poets and scientists share a vision in solitude that unites them. But what about the multitudes beneath those with cosmic visions? There is where the two cultures are found; there is where the tensions and the polarities exist.

Technology has conquered the world. We live in a technological age, not in an artistic one. We have stopped even thinking about our supreme technology and accept it implicitly. Our very language has changed. We no longer talk about creation; we talk about production. We no longer accept a phenomenon, but we measure it and verify it. The whole national effort is expressed numerically, if gracelessly, in that all-encompassing measurement, the Gross National Product. The inequality of citizenship between the artistic world and the larger legions of technology is a fundamental fact of daily life. Notice what happens when technology fails and produces malfunctioning automobiles, pollution or an exploding spacecraft. The poets and the reformers are not asked what to do to clean the rivers or repair the cars. The ultimate irony is that when technology or its application fails, more technology will be demanded.

In respect to priorities, a revolt against technology has occurred, and many scientists have been condemned because of their technological work. Conceivably the profits of technologically based industries are now in danger of disappearing, and the budgets of colleges of technolgy are not balanced. Will the flow of funds now be diverted to the novelist or to the poet? Funds will be channelled to still other technologists, especially doctors who have an immediate relationship to their human patients and who can repair the human being quickly, immediately, and visibly.

There is a reverse side, and a very melancholic one, to the victory of technology in our society. It is the defeat of the artist who, if not defeated entirely, is then in a state of defeatism most certainly. What does one do today if he is a novelist? Lord Snow's reference to T. S. Eliot's immortal words—"This is how the world ends, not with a bang but a whimper"—gives these lines a new meaning. The "world" is a literary world, the whimpering is the literary demise. Our writers have retreated. Some have sought equality with their scientific colleagues by becoming more and more poetic and more and more obscure. Now they, like scientists, are unintelligible to the public. Still others like Snow himself have chosen to write in a way he refers to as a "closed political system." That is his subject and his description of it. In the annals of literary criticism in college English departments, he is called a charlatan, not a writer. For the writer there is nothing left but greatness itself. The only thing that he can still confront with hope—as Dostoevsky did before him—is the empty page. He is not building on knowledge; he is not pushing forward to frontiers of knowledge; he is just facing the cruel, blank page. If he has read Dostoevsky, the best thing he can do is imitate him, and that is at the same time the worst thing he can do. The novelist has been submerged in our culture. The two cultures have become unequal, tilted, and unbalanced. The most melancholic comment of all is that in these first two papers, no one has mentioned this simple fact.

RESPONSE

George V. Cook

When I first gave the paper upon which my contribution to this book is based, a discussion of my address followed at the meeting between members of the faculty and the Western Electric executives at the University of Vermont. The audience asked three questions which were answered by Professors Hilberg and Felt and myself. The questions and answers serve as a commentary on my paper and are followed here by a brief summary statement by myself:

Question: Much has been written since C. P. Snow gave his famous lecture on "two cultures." His hypothesis has been largely discredited—even dismissed—as a far too simplistic statement that does not reflect the real world. There are, of course, many cultures, including the "third culture," which is also a simplistic hypothesis that will not, one hopes, tyranize us for another decade. In view of the various kinds of cultures, what do you think is the real reason that the various sectors of our society are unable to communicate with one another?

George V. Cook: I suppose this type of difficulty has always existed and will always exist. I think that an interesting approach to this problem today, however, lies in the need for a closer examination of the normative attitudes which are emerging versus the utilitarian attitudes that have dominated our society for many years. Daniel P. Moynihan, for one, has defined this new division and pointed out that it cuts across many different cultures. Joseph Schumpeter also anticipated this some years ago. We also need to recognize that rationality is not the sole possession of the scientific community and that rational people sometimes behave irrationally. As a starting point, we all need to recognize our own involvement in the very motives we criticize in others.

Raul Hilberg: We all know that scientists are rational. What are writers? Those who labor with technology should become acquainted with the meager efforts of those who must handle words as model building. The novelist is building models, hypothetical constructs of microcosmic phenomena, of both the internalizing and externalizing stimuli and his own responses. The rational-irrational dichotomy is useful, primarily because it represents a very fundamental misunderstanding of what poets or novelists are doing. They are not irrational. They are dealing with reality in a way that no other discipline is capable of dealing with it.

Mr. Cook's comment about normative and presumably purely descriptive work makes a lot of sense as a synthesis. A lawyer is a kind of normative person. He tells a client what he can and cannot do, depending on whether the client wants to win a case or not. The epistemology of law is beautiful, because there is nothing, so far as I understand law, that is undiscoverable about the facts. They exist, and if we do not know them, it is our failing. They are inherently knowable, and there is always a verdict and resolution, whether a case is won or lost. The legal process is game theory; it is an anthropomorphism. Yet descriptions and evidence are never complete, and we can never know everything. We must always grope, because we do not have an epistemology which will yield certain, objective answers. Normative and purely descriptive work is never complete. Often we do not even know which facts are relevant. But we forge ahead with a great synthesis, nevertheless. It is called a system, a system which incorporates personification on the one hand and description on the other. It behaves. No wonder this is appealing for a lawyer in a technological society. It is his own synthesis, but it is one which renders even more mechanical that which heretofore has been thought to be human.

Question: Unless we have a philosophy which provides priorities to operate the system analysis or to guide the application of modern technology, there seems little hope of really providing long-ranged solutions to our current problems. Would you comment on the kinds of priorities needed and your philosophy towards them?

Jeremy P. Felt: A philosophy concerning priorities is needed. In terms of the environmental crusade, I think for example of the housewife who feels as if she is making a great contribution to ecology by using white instead of colored paper napkins. Without a philosophy or a clear set of goals that housewife along with the general population has allowed herself to get hooked into a kind of symbolic manipulation of ecology. Some well-known people are implicitly urging this kind of manipulation. For example, David Rockefeller a few years ago stated that ecological clean-up is the responsibility of each one of us. His statement is one of those rather universal propositions that everyone agrees with, and as a result housewives devote themselves to using white instead of colored napkins. What is missing (like the cry of the 1880s when certain politicians said what is needed to clean society up is the election of honest men to office) is the responsibility of the larger economic institutions of society to really do something based on widespread agreement. In my example, the government and industry might agree not to manufacture colored napkins.

Question: Could not solutions for the critical problems threatening human survival best be found by rational inquiry and the adoption of the results?

George V. Cook: One experience I recall about the beginning of my legal education concerned the search for truth. We will always have situations in which very rational men will disagree. On my first day in law school, we had no limitation on class hours. After three and a half hours of discussing a case, one student rose, stood on his chair, and waved his hand. The student shouted from his desk that he thought the plaintiff won the case. The instructor asked for the student's reasons, and he replied, "Because the Supreme Court of the United States said so." The old instructor looked at him and said, "And you believe them?" We are in the era where we need to recognize that rational people can disagree.

Jeremy P. Felt: To state it as antagonistically as possible, the problem we confront is that rational arguments are rooted in irrational assumptions and that people who administer the quality of our lives are confused. Our leaders wear a veneer of rational action as they progressively grab various banners and hold them up as a kind of rational patina for what they are doing. What one calls rational inquiry is simply another banner to be waved in the air. Suppose we were to discover, for example, that a society had existed between 9000 and 5000 B.C. in which the most rational form of behavior and environment had been devised. Can you imagine that we would begin to try to duplicate that kind of rational living in our own society? Because we have pledged allegiance to what we imagine is progress-through-technology, a very great effort would be made to denounce this earlier, rational society as impractical and unreasonable. On a serious level, there is an addiction to the notion of growth and progress as a new creed. That addiction may make us unable to accept the results of rational inquiry.

Raul Hilberg: Once upon a time I went to a meeting of political scientists and wandered into a panel of graduate students reading their papers. I listened to a paper on the successful efforts by agencies in state legislature hearings to acquire their respective budgets. As I listened, I became increasingly interested. The student presenting the paper had discovered a very fundamental thing. Over a period of ten or twenty years, he found that that agency which would do a little better in the competition for funds would become emboldened to ask for even more money. Those agencies which did not do so well in initial competitions tended to ask for a little less and became a little more defensive each year. Now ostensibly hearings before legislatures have a certain amount of rationality in that money is requested for specific tasks and needs. Yet this young researcher was able to show that the primary factor in distributing scarce resources was success or failure itself. Once an agency acquired certain power, it would inevitably get more power. Rational arguments will be even more rational. And once an agency has failed to obtain funds, it will inevitably get less money, and what was not so convincing last year will be even less convincing this year. Irrational arguments will be even more irrational.

In conclusion, I did not mean to imply, as Professor Felt suggested, that in dealing with tensions, a ruling principle ought to be to place external limits on the extent of disagreement. On the contrary, it is my thesis that the frictions brought about by disagreements are central to civilization; but we should at least, in the process, have sympathy for the other man's vocabulary. We need somehow to address ourselves to the defeatism to which Professor Hilberg alludes and to bring some optimism to our imperfect world. The whole argument toward the system's approach is nothing more than an argument that other people's understandings and other people's disciplines may provide many different perspectives. I have no substantive solutions. My solution is really Maxwell's "demon," which is not a predetermined means of solution but rather suggests that individuals can, by being "demons," make substantial contributions to the identification of problems and their solution. A humorous part of one of the Maxwell biographies was an interchange with Lord Kelvin about the invention of the telephone. When Maxwell was told about the invention of the telephone, he said, "This is marvelous, but there is absolutely nothing here that we didn't know already." The irony of this statement is a fitting conclusion for my comments.

BIBLIOGRAPHY

Fuller, Edward, "Snow-Leavis Affair," *New York Times Book Review*, 22 April 1962, pp. 24 f.

Green, Martin, "A Literary Defence of the 'Two Cultures' ", *Critical Quarterly*, IV (1962), pp. 155ff., and *Kenyon Review*, XXIV (1962), pp. 731 ff.

Green, Martin, *Science and The Shabby Curate of Poetry, Essays About the Two Cultures*, (W. W. Norton, Inc., New York, 1965).

Holton, Gerald (ed.), *Science and Culture*, Beacon paperback, Boston, 1967.

Leavis, F. R. and Michael Yudkin, *Two Cultures? The Significance of C. P. Snow*, (New York, 1963).

Leavis, F. R., "The Significance of C. P. Snow," *Spectator*, 9 March 1962, pp. 297 ff.

Snow, C. P., "The Two Cultures," *New Statesman*, 6 October 1956.

Snow, C. P., *The Two Cultures and The Scientific Revolution*, (Cambridge University Press, 1959).

Snow, C. P., *The Two Cultures and a Second Look*, (Cambridge University Press, 1964). Paperback edition (Cambridge University Press, 1969).

Trilling, Lionel, "Science, Literature and Culture, A Comment on the Leavis Snow Controversy," *Commentary*, 33 (1962) pp. 461, ff.

SCIENCE
AS
A
CREATIVE
ART

INTRODUCTION

The topics of the first and second sections furnish a context for "Science as a Creative Art." Professor Williams asserted in his paper that there was an unbridgeable gap between the rare scientist, or person of broad vision and deep commitment to a personally held truth, and the problem-solvers, the practitioners of science and technology. Historically, the creative genius provided the direction for the problem-solvers to follow. George V. Cook chose to give brief mention of the role of the individual creative genius, because it was intangible and undefinable. He stressed, instead, the need to improve modern team research and systems engineering by providing greater interdisciplinary fertilization. Finding Lord Snow's distinction of two cultures insufficient, he stressed the "infinity of cultures" contained by our society. We have a pressing need, he argued, to improve communications between people and to try to eliminate myopic insularity, so that we may have an optimum situation for problem-solvers to seek solutions to our many obvious problems. Professor Steffens expands upon the themes of these authors by accepting the arguments and distinctions of L. Pearce Williams' paper and by rejecting many of the assumptions and orientations in George Cook's paper.

Professor Henry John Steffens discusses the twentieth century misconception of science, and the often-neglected similarities between scientific and artistic creativity. He argues that popular conceptions of science too often confuse science with technology and call both "science." He suggests that this is a natural error, stemming from the increasingly close working relationship between science and technology which began after the 1850's when scientific achievement became the obvious and potent source for the most dramatic technological changes. Science was linked to progress, and progress was determined by new things produced and by societal changes. Professor Steffens points out, however, that what was neglected was that the science which aided the production of the internal combustion engine, the chemical industry, and wireless communication was the science of the nineteenth century. Twentieth century science moved onto new and different understandings of the world, while the twentieth century public remained largely oblivious to the new world of the modern physical sciences. The public conception of science remained largely the classical nineteenth century model of objective statement, methodological techniques, and the confidence that the truth was being revealed.

Professor Steffens asserts that "science is not simply a disinterested, objective, and mathematical endeavor guided by an explicit set of methodological rules." Science cannot, in any way, guarantee or even imply material progress, nor can it offer the assurance that it will be able to provide eventual solutions to all questions posed. On the contrary, intuition and imagination play a central role in science. "The most revolutionary discoveries and the most profound changes in the direction of science have usually occurred on the basis of old, well known information and facts." The profound changes which occurred in science as a result of the work of men like Galileo and Newton were all dependent upon new understandings based upon known facts. They were the results of the exceptional intuitive powers of the individual scientists.

The existence of scientific intuition is demonstrated historically, Professor Steffens claims, but the way in which this intuition comes into effect in the process of scientific discovery is quite a different matter. There can be no rules or laws for scientific intuition, for creativity is outside the context of scientific and technological methodology. Professor Steffens presents several examples of how an aesthetic appreciation of the beauty and simplicity of new discoveries has frequently accompanied the most fundamental scientific advances. He suggests that such aesthetic appreciation is closely linked to the scientist's basic philosophy, to his view of the world, and to the order he perceives. Thus, scientific creativity is closely akin to creativity in the arts; artistic and scientific creativity seldom pay much attention to methodological strictures and demands.

Professor Steffens concludes by calling for the revival of a serious concern for metaphysical questions. He argues that a concern for such questions is important both for the broadening of the basis of our technological capabilities and for fundamental scientific change. Our current emphasis on behavioralism in the life and social sciences and on positivism in the physical sciences does not lend itself to these purposes. These formalisms, he asserts, can now be recognized as limited and we are faced with a choice between agnosticism or a belief in the emergence of a new natural philosophy. Professor Steffens does not address the question of the adoption of a new natural philosophy by society. Jeremy P. Felt has raised the question earlier in this book of who shall decide which technological strategy should be adopted. If Professor Steffens is correct, there may be no effective solutions to complex technological problems without a broader understanding of the context within which these problems occur. In this situation, more technology to solve technological problems is futile and generates more problems than it solves.

Although in substantial agreement with Professor Steffens, Professor Albert Crowell emphasizes the dual nature of scientific activity: science as an information-collector and problem-solver and science as a creative art. He suggests that although men of the stature of Einstein and Newton obviously contributed to science, so did the men who worked industriously at small problems. He also warns that if a new natural philosophy emerges, either by revolution or by evolution, it will represent even greater potential power in the hands of some people. This power can lead to either a better or worse way of life in the world.

William Banton, in disagreeing with Professor Steffens, opens his commentary by stating that "knowledge for knowledge's sake is sterile. Applied technology has created the world in which we live." He points to what he views as a concurrence between the rise of the technical university curriculum and the Industrial Revolution which in turn created the demand both for engineers or technologists and for science and technology to cease being separate endeavors. He questions whether this loss of distinction between science and technology has proved harmful or beneficial to mankind. The problem to consider, he feels, is whether this development is "bad," in view of the fact that America has the highest standard of living in the history of man and at the same time has the highest level of pollution known to man.

Mr. Banton also takes issue with the question of the nature of creativity which Professor Steffens presented. He suggests that creativity is not a simple functional relationship within the brain. It must include the element of "idea association ability," since the ability to associate ideas produces a tremendous difference in the creativeness of individuals. Mathematically, "creavity increases as the log normal distribution." A theory of idea association does provide, he argues, a plausible theory of creativity and of what causes an Einstein. He adds that idea association is taught in some colleges of technology in courses entitled, "Creative Engineering," or in the closely related field of "Value Engineering."

In conclusion, Mr. Banton cautions against confusing necessary with sufficient conditions in the consideration of science policy. Strong basic science is a necessary condition for a strong economy, a livable environment, and a tolerable society, but it is not a sufficient condition. "It is this test of sufficiency to which we are all addressing our current activities."

Robert Clagett, while agreeing with Professor Steffens' suggested changes in university education and in the ways scientists view their profession, also raises several points of objection. He suggests that it is not clear how to design an educational system which would create genius, but that it would be difficult to design one which would prevent a genius from making significant scientific contributions. Mr. Clagett also objects to an educational system structured to foster the few true men of science that Professor Williams isolated in his paper.

Finally, Mr. Clagett addresses himself to the problem of man's necessary involvement in everyday solutions to current problems. He argues that industry cannot support men who do not have practical concerns, or at least direct expectation of practical applicability, because an industry which allows such expenditure would be at a competitive disadvantage, its products would be overpriced, and it would have decreased profitability. Moreover, government sponsorship of research that has no foreseeable practical application would have the same problems, because of concern for results and for taxpayers' money. The university, Mr. Clagett concludes, must support individuals with interests beyond current problem-solving and application, with assistance from both industry and government.

Mr. Clagett's commentary raises the issues of industrial, governmental, and university support for basic research. For the industrial environment, basic research is usually viewed as a high risk expense and is often unpopular with stockholders despite the potential for high reward at a later date. In some cases, basic research may even be imperative to the ultimate survival of a particular industry. Why, on the other hand, industry can justify its support for basic research in the government or in a university is not made clear.

SCIENCE
AS
A
CREATIVE
ART

Henry John Steffens
Associate Professor of History
University of Vermont

Twentieth century Western man seems to have failed to appreciate the nature of science. Recent world events and rapid technological developments have served to merge the two fundamentally different human endeavors of science and technology into one popular conception now labeled "science." This serious mistake becomes ironical when we remember that all societies have produced technology, but only Western civilization succeeded in producing that special form of intellectual creativity called science.

The origins of our misunderstanding of science can be discerned if we consider modern European history. The most obvious source of our error has been the remarkable success of the West's technological development during the Industrial Revolutions of the nineteenth and twentieth centuries. Technology and science became irrevocably intertwined in Western conceptions after the mid-nineteenth century. Before the 1850's and 1860's, technology and science were both developing very rapidly, but with little mutual interdependency. The engineers and manufacturers had their own set of specialized problems which emerged from the rapid advance of manufacturing techniques and improvements in transportation and mining. The creativity of those technologists was sufficient to enable them, for example, to improve the steam engine and its applications with little or no help from science. The scientists, or natural philosophers, on the other hand, devoted themselves to the study of the new wave theory of light, to electric and magnetic phenomena, to the new theory of heat, to electrochemistry, and to the new conceptions of a sidereal universe. In short, the scientist before the 1850's had little to say that would be useful to the technologist, despite his unabashed enthusiasm for the world-view emerging from his work.

But after the mid-century, this basically non-useful relationship changed dramatically with the development of the conservation of energy principle and the concept of the electro-magnetic field. James Prescott Joule's experimental work on the mechanical equivalent of heat was combined with basic philosophical notions about the correlation of the forces of nature and with newly developed mathematical techniques to produce a new understanding of the physical world in terms of energy. The earliest possessors of this new understanding—Lord Kelvin, Hermann von Helmholtz, and William Rankine—expressed this world-view, in part, in the laws of thermodynamics. Once these laws were expressed and explained, the natural philosopher was in the unusual position of being the well-spring of ideas for technological improvements. The steam engine and the new internal combustion engines were direct beneficiaries of this new understanding of the physical world, and a new kinetic theory of heat also followed rapidly.

The same changes, in general, occurred in the area of electro-magnetic field theory. Michael Faraday's clear experiments and profound insights into electrical and magnetic phenomena were not lost upon the young James Clerk Maxwell. His understanding of what Faraday meant by his concept of the electro-magnetic field and his enthusiasm and ability in mathematics served him in devising his famous "Maxwell's equations" of the electro-magnetic field. These equations and the understanding of their use, served as the foundation for the obvious and rapid development of the uses of electricity and magnetism in the late nineteenth century.

The world after 1870, then, was literally a different world, both conceptually and in terms of rapid technological developments which dramatically changed men's lives. It was this startling change which produced our conceptual problems. Science became the clear source of the most dramatic directions for technological improvements. Electric power, in all its forms, the new chemical industry, internal combustion engines, wireless communications—all owed their beginnings to nineteenth century scientific developments. The link between science, technology and the production of equipment and devices was not only forged; men concentrated upon it. What was forgotten was that science quickly moved into new areas of investigation, unconcerned with the increasing applications produced and societal changes made.

The public view of science emerging from the process of change was obvious by the beginning of the twentieth century. Science became linked to a scientific method and scientific procedures which served as the basis for applied science and technological progress toward specified goals. Being "scientific" rapidly came to mean doing those things which produced results most quickly, be they new things such as new chemicals, or new answers to specific, though sometimes very complex problems. Science meant progress, and progress was determined by new things produced and changes made in society.

This emerging public view of science was reinforced in several ways. Powerful support came from the obviously close relationship between developments in biology, biochemistry and medicine and their immediate salutary effects on society. The recognition of the germ theory of disease, discoveries of chemicals to aid in fighting disease and a better

understanding of body chemistry led to tremendous improvements in the ability to supply health care by the beginning of
the new century. Here science was linked to direct societal
effect in a most personal and immediately perceivable way.

The remarkable success of the "team research approach"
to problems also reinforced this public image of science,
especially after World War II. The image of the lone inventor
attempting new solutions by trial and error was replaced by
that of the research team, a group of people with many types
of "expertise" to use the most recent "state of the art" in
finding the solution to a specified problem. It was clearly
assumed that the methods such a team used to find their
answers were in strict agreement with the scientific method.
In fact, it was generally assumed that success of the team
followed with a rapidity which was in direct proportion to
the team's degree of adherence to the scientific method.

Modern trends in philosophy and the social sciences have
lent support to this public image of science. Although a
complete discussion of trends in these fields is not intended,
it should be stated that the concepts of scientific method and
science as an empirical, objective process have been emphasized by the dominant trends in modern philosophy and
social science. Many philosophers have striven for the logical
rigor and analytical clarity of scientific laws in their philosophy. Most social scientists consider themselves to have
succeeded or failed in their research by measuring how
closely their endeavors have been conducted by "scientific
methods" to turn up empirically produced, general statements about society or human behavior. The public conception of science has served as an important model and stimulus
for a great deal of twentieth century intellectual life. Unfortunately, this model was derived by extension from views
obtained from the nineteenth century merging of science and
technology. This model has very little to do with science, but
a great deal to do with applied science and scientific technology.

Science is simply not a disinterested, objective, pure,

mathematical, non-qualitative endeavor guided by an explicit and rigorous set of methodological rules. It never has been and it never will be. Science can not guarantee, or even imply material progress, nor can it offer the assurance that it will be able to provide eventual solutions to all questions posed. In the public view, science has been linked to change and to the control of nature. This notion is so incorrect that several lines of argument might be needed to fully illustrate the error.

Magic has failed in Western tradition because its goal was the powerful control of nature. Science has succeeded as an important intellectual tradition because it has always been searching for an understanding of nature for its own sake. Science, by producing an understanding of nature, enabled Western men to put this understanding to obvious and somewhat amazing practical application. But the understanding preceeded the application; the power came from the understanding of nature, not from the attempt to gain ascendancy over nature.

Magic produced no understanding of the physical world, simply knowledge of various effects. It sought to apply its own well-defined methods and techniques to the solution of specified problems and to impose control over nature. It failed. It is certainly no accident that about technology we so often hear today the statement: "it's magic." Failure to understand has been the prime characteristic of magic. Since technology provides no understanding of the physical world, only a knowledge of processes and effects, it has become the natural recipient of the revived term, "magic." Our ability to open garage doors at a distance or by turning a knob to produce a color picture, complete with sound is nothing short of magical. It remains magic unless we dispell the mystery by learning to understand nature in terms of electro-magnetic theory.

A reference to a Greek story may be instructive. The story is about Orpheus and his ability to play the lyre. Orpheus had such a deep understanding of music and such an ability to reveal his understanding by the beautiful music he played, that he could tame savage beasts by playing the lyre. It would be ridiculous to assume that Orpheus developed his gift of music by setting out to become a lion-tamer! (1)

Science has always been the search for an understanding of nature. It is a search for the hidden unity behind the confusing multiplicity of events. It is a search which recognizes and enhances the importance of imagination and intuition. It is an endeavor which has provided an opportunity for Western man to display his most profound creativity.

Science is not simply a collection of well-ordered facts. If science consisted of facts, and scientific theories emerged from observations and experiments, then to understand nature we would only need to find and record data as carefully and as completely as possible. All the scientist would need do is look to his data and, using the appropriate mathematics as a tool, work the data into scientific statements and conclusions. This approach sounds persuasive, and it serves to represent a large part of the popular view of how a scientist works. The fallacy of this orientation may perhaps best be revealed by a fable, rather than by detailed logical arguments concerning the problems of induction.

The fable was created by Sir Karl Popper. A person wanted to become a scientist and devote his life to science. Proceeding on the assumption that science is based on observed facts, he would record all his observations as accurately and carefully as possible. During the course of a diligent and full life, he might, after thirty or forty years fill several hundred notebooks with recordings of temperature, humidity, cosmic radiation levels, shifts in the earth's magnetic pole, movements of the polar ice caps, stock market fluctuations, length of ladies' hemlines, air pollution levels; in short, all the data he observed would be noted. In his will, of course, he would bequeath his notebooks to a scientific academy, content in the knowledge that his life's contribution to science would not go unappreciated. Of course, no scientist would even want to look at the notebooks. All they contained would be a random jumble of meaningless data. The facts, in science, as

everywhere else, certainly do not speak for themselves!
Science attempts to find meaning and to discover the order
behind our observations and experience. Scientific theories
do not emerge from the data, but quite the contrary are free
creations of the human mind. Man is a participant in the
reality he creates. As J. Bronowski describes it:

> Reality is not an exhibit for man's inspection,
> labeled 'Do not touch.' There are no appearances
> to be photographed, no experiences to be copied,
> in which we do not take part. Science, like art, is
> not a copy of nature but a re-creation of her. We
> re-make nature by the art of discovery, in the
> poem or in the theorem. And the great poem and
> the deep theorem are new to every reader, and yet
> are his own experiences, because he himself re-
> creates them. They are marks of unity in variety;
> and in the instant when the mind seizes this for
> itself, in art or in science, the heart misses a beat.(2)

Albert Einstein agreed that science must be the act of
discovery of the creative mind. "Science is not just a
collection of laws, a catalogue of unrelated facts," he wrote.
"It is a creation of the human mind, with its freely invented
ideas and concepts. Physical theories try to form a picture of
reality and to establish its connection with the wide world of
sense impressions. Thus the only justification for our mental
structures is whether and in what way our theories form such
a link." (3)

As Einstein himself, and Heisenberg and Schrödinger, Bohr
and Dirac, Rutherford and Pauli, demonstrated so clearly in
the first decades of our century, the science thus freely
invented is not the true expression of complete and everlast-
ing reality. Scientific theories are certainly subject to change,
even though they can be linked to our world of sense
impressions. Far from being permanent, it is remarkable that
they operate for as long as they do. Considering the changes

in our world-view wrought by modern physics, it is difficult
to understand why the science created in the mind of Sir
Isaac Newton succeeded as well as it did.

Intuition plays a central role in science, for it serves as
both a guide to doing science and as a source for new
orientations and understanding. The importance of intuition
to science is clear to the student of the history of science.
The most revolutionary discoveries and the most profound
changes in the direction of science have usually occurred on
the basis of old well known information and facts. Revolutions seldom occur in response to new facts, but usually
occur because of a re-interpretation of facts long known and
familiar. Revolution based upon factual re-interpretation was
true of Copernicus and his new astronomy; it was true of
Galileo and his new laws of dynamics; it was true of Kepler
and his laws of the motions of the planets; it was true of
Newton and his work in optics; it was true of Lavosier and
his chemical revolution; of Dalton and his atomic theory; of
Darwin and his natural selection; of DeBroglie and Schrödinger in their new interpretation of wave phenomena. All of
these new understandings were the results of new inferences
from known facts based upon the exceptional intuitive
powers of the individual scientist. The existence of scientific
intuition is demonstrated historically beyond question.

The way in which intuition comes into play in the process
of scientific discovery is quite a different matter. Plato made
it clear in the *Meno* that the process of discovery of new
understanding is very difficult to fathom. Michael Polanyi uses
the example of Plato to point out that:

> to search for the solution of a problem is an
> absurdity. For either you know what you are
> looking for, and then there is no problem; or you
> do not know what you are looking for, and then
> you are not looking for anything and cannot
> expect to find anything. If science is the under-
> standing of interesting shapes in nature, how does

this understanding come about? How can we tell what things not yet understood are capable of being understood? (4)

The scientist is certainly not left awash in a sea of intuitions about the physical world, but just as certainly no well-defined rules appear for the use of his intuition. As Michael Polanyi states the case:

> Admittedly, these are rules which give valuable guidance to scientific discovery, but they are mere-ly *rules of art*. The application of rules must always rely ultimately on acts not determined by rule. Such acts may be fairly obvious, in which case the rule is said to be precise. But to produce an object by following a precise prescription is a process of manufacture and not the creation of a work of art. And likewise, to acquire new knowledge by a prescribed manipulation is to make a survey and not a discovery. The rules of scientific enquiry leave their own application wide open, to be decided by the scientist's judgment. This is his major function. (5)

Albert Einstein, characteristically, has put this problem of scientific intuition in more immediately perceivable terms. He enjoyed reading a fine mystery and drew a comparison of the similarities and differences between the detective and the scientist.

> In nearly every detective novel since the admirable stories of Conan Doyle there comes a time where the investigator has collected all the facts he needs for at least some phase of his problem. These facts often seem quite strange, incoherent, and wholly unrelated. The great detective, however, realizes that no further investigation is needed at the

moment, and that only pure thinking will lead to a correlation of the facts collected. So he plays his violin, or lounges in his armchair enjoying a pipe, when suddenly, by Jove, he has it! Not only does he have an explanation for the clue at hand, but he knows that certain other events must have happened. Since he knows exactly where to look for it, he may go out, if he likes, to collect further confirmation for his theory.

The scientist reading the book of nature, if we may be allowed to repeat the trite phrase, must find the solution for himself, for he cannot, as impatient readers of other stories do, turn to the end of the book. In our case the reader is also the investigator, seeking to explain, at least in part, the relation of events to their rich context. To obtain even a partial solution the scientist must collect the unordered facts available and make them coherent and understandable by creative thought. (6)

There can be no rules or laws for scientific intuition, and it is impossible to define exactly where this intuition originates and how it proceeds. This mystery, of course, is the central and most serious problem confronting the historian of ideas. It is however possible to describe the intellectual milieu within which this intuition arises with careful accuracy. Two of the most persistent ingredients in the background of important scientific discoveries are the ideas of beauty and the belief in the inner harmony of the world.

An aesthetic appreciation of the beauty and simplicity of new discoveries has frequently accompanied the most fundamental scientific advance. The element of aesthetics was most frequently elicited by men who had mathematical abilities. Copernicus informed us in the preface to his great work in astronomy that he was disturbed to find that the Ptolemaic astronomers had produced such a cumbersome scheme of the universe with their epicycles, equant points and eccentric circles, that he could only call that system a "monster." He sought to simplify the scheme of the universe to make it more pleasing and to make it conform more closely to the beautiful universe God must have created. After all, God would certainly not create anything as ugly as Ptolemaic astronomy!

P. A. M. Dirac was moved by similar sensibilities as he approached the problems of modern quantum physics. "All great discoveries are beautiful," Dirac wrote, "It is more important to have beauty in one's equations than to have them fit experiment." (7)

In 1940, as he reviewed his long productive life in mathematics, G. H. Hardy, the English mathematician, wrote:

> The mathematician's patterns, like the painter's or the poet's, must be *beautiful*; the ideas, like the colours or the words, must fit together in a harmonious way. Beauty is the first test: there is no permanent place in the world for ugly mathematics . . . It may be very hard to *define* mathematical beauty, but that is just as true of beauty of any kind—we may not know quite what we mean by a beautiful poem, but that does not prevent us from recognizing one when we read it. (8)

The emphasis on beauty can be related to the way men work in science by a profound definition of the concept of beauty. This definition was first espoused by the Pythagoreans but received understanding and support from the poet, Samuel Taylor Coleridge, in the beginning of the nineteenth century. Coleridge was deeply interested in science and had friends at the Royal Institution in London. He lectured at the Royal Institution and was a friend of Sir Humphrey Davy. Coleridge defined beauty as "unity in variety." In his essays, *On the Principles of Genial Criticism*, he referred to the Pythagoreans and wrote: "The safest definition, then, of

Beauty, as well as the oldest, is that of Phythagoras: THE REDUCTION OF MANY TO ONE." (9)

The acceptance of this definition of beauty conveys an explicit orientation toward science. Science becomes the search to understand the unity behind the vast variety of experiences confronted by our senses. Science and consequently poetry, painting and all the arts become the same kind of search, the search for unity in variety. Each area of intellectual endeavor searches for the hidden likeness and unity behind the appearances of disorder and change offered to man at first glance by the world of the senses. Scientific theorems become those statements of unity which enable us to understand the relationship between previously unrelated phenomena in the physical world.

James Clerk Maxwell's equations offer both an example and a powerful support for the view of science as an attempt to understand unity. Maxwell was able to express Faraday's understanding of the relationship between electricity and magnetism in terms of mathematical equations. Maxwell's formulation allowed him to understand electro-magnetic phenomena in such a way that he could relate the separate development of the wave theory of light to electricity and magnetism. Three fields of inquiry were united by the concept of the electro-magnetic wave. The relationships between electric and magnetic fields and electro-magnetic waves were explicitly revealed by Maxwell. Heinrich Hertz took Maxwell's revelation of unity seriously. From what he knew from his understanding of Maxwell's work, he attempted to observe what should be observable, the electro-magnetic wave. Hertz undertook to look for the electro-magnetic wave because of his new understanding of electro-magnetic phenomena. He looked for the waves and found them.

Albert Einstein also believed that science was deeply concerned with the search for unity and the inner harmony of the world. His very personal belief served to direct his scientific work. The last paragraph in his work, *The Evolution of Physics*, portrays his attitude most poignantly:

> With the help of physical theories we try to find our way through the maze of observed facts, to order and understand the world of sense impressions. We want the observed facts to follow logically from our concept of reality. Without the belief that it is possible to grasp the reality with our theoretical constructions, without the belief in the inner harmony of our world, there could be no science. This belief is and always will remain the fundamental motive for all scientific creation. Throughout all our efforts, in every dramatic struggle between old and new views, we recognize the eternal longing for understanding, the ever-firm belief in the harmony of our world, continually strengthened by the increasing obstacles to comprehension. (10)

Einstein's personal definition of science, as well as much of the portrayal of science I have developed, has been questioned by the usual interpretation of the development of quantum physics. The issues are extremely complex, but the central concern may be presented as the question of the applicability of the principle of causality. The success of modern quantum physics has been interpreted to indicate the necessity of abandoning a conception of nature in which there are detailed causal relationships and which may no longer be considered in terms of "inner harmonies." Einstein's orientation toward nature and his belief in the ability of science to provide an understanding of the harmony of the world were viewed as old-fashioned by physicists in the 1930's and 1940's. His beliefs were considered to be the peculiarities of a man growing old.

The interpreters of the philosophical implications of quantum physics were concerned with the probabilistic statements of events, with indeterminacy and with the changed notion of causality which they believed was required by quantum theory. Erwin Schrödinger has provided a statement of the situation which is very useful for our purposes. In his address given at his inauguration to membership in the Prussian Academy of Science, in July, 1929, Schrödinger discussed the difference between the traditional notion of the concept of probability and the new conception derivable from quantum physics. In classical physics, individual events were treated statistically for convenience. Ludwig Boltzmann's statistical treatment of the kinetic theory of gases, for example, represented gas molecules as parts of statistical samples of a gas. The individual gas molecule was not considered, because there was no need to consider it and it was much more convenient not to. In principle, no reason existed to believe that the individual molecule could not be treated; in fact every reason to believe, in the context of classical physics existed that the individual could be represented, if required. Referring to this situation, and the situation of all material bodies composed of atoms, Schrödinger said:

> Uncertainty in this case arises only from the practical impossibility of determining the initial state of a body composed of billions of atoms. Today, however, the doubt as to whether the processes of nature are absolutely determined is of quite a different character. The difficulty of ascertaining the initial state is supposed to affect not merely a complicated system, but even a single atom or molecule. Since what is by no possible means observable does not exist for the physicist as a physicist, the meaning clearly is, that not even the elementary system is so exactly defined as to let it react to a definite influence by a definite behavior. (11)

This passage, which agrees with what is now referred to as the "Copenhagen" or "Usual Interpretation" of quantum theory, provides the clue to where this interpretation has been incorrect in its emphasis. It was artificial and inappropriate to isolate "the physicist as a physicist." A specialized approach to a creative intellectual endeavor was fundamentally incorrect, but it was quite in keeping with the development of our twentieth century conception of science. On the level of real creativity and originality, science has rarely been done as "science," but almost always as "natural philosophy." The range of creative ideas has always been greater than the proscriptions of principles of observability and non-observability. The usual interpretation of quantum physics has attempted to force an either/or choice between causality and indeterminacy. The argument is as follows: Quantum physics, with the aid of logical positivism and analytical philosophy, has assigned very precise limits to the knowledge derivable from sense experience; or, epistemology has been developed with extreme sophistication in the twentieth century. Unfortunately, this epistemology has a great deal to do with what we can logically claim to "know" about the physical world, but rather little to do with science as natural philosophy.

The creative natural philosopher, seeking a new understanding of nature, has never paid much attention to the strictures and definitions of an a posteriori interpretation of what happened, based upon an after-the-fact analysis of what the new developments in physics "really" mean. Science or, as I prefer, natural philosophy has a way of moving out from under the philosophical constructions raised about it.

The usual interpretation of quantum physics has served to stress many extremely important concepts and implications which arise from the new physics. This interpretation has served to clarify these points and they must, of course, be well taken. Perhaps most important for our conception of modern science is the necessity of eliminating the search for models of the phenomena discussed by the quantum formalism, based upon concepts taken from the macroscopic world of classical physics. It has been made abundantly clear that the standard notions, such as position and momentum, derived from the physics developed in the nineteenth century, cannot be forced down to the level of electrons and photons. The terms of the standard, determined, mechanical interpretation of nature simply can not be applied to the realm of nature where Planck's constant becomes significant. What has however been demonstrated is the inappropriateness of this approach to knowledge, *not* that the world of the very small is indetermined or acausal! As Galileo perceived at the beginnings of modern science, nature obeys laws, whether man is capable of grasping those laws or not.

The modern stress on epistemology has allowed an inappropriate extension to take place. The epistemological problems presented by quantum physics have been converted into "in principle" statements. It is claimed that the world is, in principle, indetermined. This "in principle" development of the challenges presented by epistemological problems raised by modern physics has resulted in the overly hasty extension of this position to include statements about what the world "is really like." Epistemology has been extended to include ontology.

Our western philosophical tradition, especially after Bishop Berkeley wrote his *Dialogues,* should have remembered that the relationship between epistemology and ontology was far from as direct as the usual interpretation of quantum physics would have us believe. As Euphranor instructed Alciphron as they both viewed a distant castle: "I would infer that the very object which you strictly and properly perceive by sight is not that thing which is several miles distant."

Erwin Schrödinger appreciated this difference in his address before the Prussian Academy in 1929. Speaking of the causal vs. the indeterminate concept of nature, he said:

> But I do not believe that in this form it will ever
> be answered. In my opinion this question does not
> involve a decision as to what the real character of a

natural happening is, but rather as to whether the one or the other predisposition of mind be the more useful and convenient one with which to approach nature . . . We can hardly imagine any experimental facts which would finally decide whether Nature is absolutely determined or is partially indetermined. The most that can be decided is whether the one or the other concept leads to the simpler and clearer survey of all the observed facts. Even this question will probably take a long time to decide. (12)

In the years after Schrödinger's address, his cautious words were neglected as philosophers, philosophers of science and many scientists opted for indeterminacy and confidently extended this concept to "the real character of a natural happening." But it has become clear that the notions of a "partially indetermined nature" are *not*, in fact, offering the most "useful and convenient" approach to nature. There can be no denying the tremendous short term success of this orientation. The remarkable achievements in the physical sciences from the 1920's to the 1970's more than attest to this orientation. But it can now be recognized that this usual interpretation of quantum theory has been worked by philosophers of science into the kind of closed system that seems no longer to offer a path to a new and deeper understanding of nature. The scope of the quantum formalism has been precisely defined and limited. There have even been mathematical proofs attesting to the completeness of the quantum mechanical assumptions.

But, science has moved out from under this formalism and interpretation. The path to a deeper understanding of nature seems now to lie in the abandonment of the "partially undetermined" world-view and toward a return to what has always been the natural philosopher's fundamental concern: the search for the unity and inner harmony of nature. This new search must be renewed, but now conducted in a "sadder, but wiser" way, with the epistemological lessons of quantum theory firmly in mind. The new natural philosopher must recognize that he will never achieve a full understanding of nature. Unlike the optimistic men of the late nineteenth century, he must be prepared to marvel that the laws he formulates about nature apply as well as they do. He must recognize the full implications of the limited uses of classical models to describe the world. The uncertainty principle gives clear information in this regard and the persistence in the use of models leads only to the formulation of inappropriate questions which have no answers in the form they are asked. He must appreciate the need for mathematical representation which is even further removed from the realm of common experience than was the formalism of classical science. These limitations change the endeavor, but certainly do not eliminate the possibility of doing natural philosophy.

The direction from which a new, creative understanding of nature will come is unpredictable, if it comes at all. Since the direction cannot be defined, the new orientation which natural philosophy will take is also unpredictable. That a new approach is needed is now clear from a number of technological as well as scientific considerations. Our understanding of nature provides the context for and the limitations of scientific technology. The "scientific method," as it is defined and accepted by technology and applied science, is tremendously effective only within the limits of understanding and interpretation achieved by natural philosophy. Scientific technology can solve well defined problems with remarkable efficiency. But, if the problem and the solution are not included within the context of our understanding, the possibility of finding the solution becomes directly related to the trial-and-error empirical approach used by the lone inventors of the nineteenth century who operated largely outside of science.

Our "team research" approach to modern problems assumes that by careful choice of team members, only those areas which appear likely areas of investigation are attacked.

These areas are dependent upon our current natural philosophy. If the current natural philosophy has nothing important to say about areas which should be selected for intensive "team effort," our whole technological approach falls back to an intensive trial-and-error, hit-or-miss attempt to find solutions. Worse still, without an understanding of nature which includes the solution to the problem we seek, we may not even be capable of recognizing the solution, if we were to stumble across it. As Louis Pasteur claimed, discovery only happens to the prepared mind.

A growing number of modern technological and scientific problems now appear to be both too complex for trial-and-error solution and outside the current capabilities of our understanding of nature. Problems of the environment, a cure for cancer, an understanding of nuclear particles and interactions, new celestial phenomena and the reconciliation of quantum theory with relativity theory seem beyond the capabilities of our current scientific formalism as it is popularly interpreted. In short, if we are to have any legitimate and well founded hope of continued advance in scientific technology to meet the increasingly complex problems presented by our modern world, we must have the help of a new understanding created by a new natural philosophy. If this understanding is not forthcoming, we must resign ourselves to the possibility of piecemeal and fortuitous technological solutions to our problems, based on chance and not on understanding.

Several possible advantages may be gained by a change in our orientation. The creation of a new natural philosophy, enabling us to perceive unities in all fields of science, might well prove capable of important unifying statements between the physical and biological sciences. The interpretation of science as natural philosophy implies that there should be a basic unity to all fields of science. All sciences should be related by the fundamental understanding of nature from various directions and points of view. Although the various fields of science seem to be different and more specialized in terms of the wealth of their specific details, they should be regarded as basically related. Rather than viewing the sciences as more and more fragmented and specialized, we might more profitably take the opposite approach and attempt to find the similarities and interrelationships of the sciences. We should not be fooled by the vast quantity of knowledge produced by each specialized field into thinking that no common areas of understanding exist. Natural philosophy, in all fields, can be viewed as the same endeavor: the creative search for unities and inner harmonies behind the world of appearances. If this orientation toward science is correct, it is conceivable that we may reach a new level of understanding of the natural world which will have important, unexpected implications for both the biological and physical sciences. The specialized collection of data in literally thousands of seemingly isolated disciplines should be viewed as necessary support activity which yields information, but little understanding. This accumulation of research, considered by the twentieth century as the essence of scientific activity, should at our stage of development be viewed as of distinctly secondary importance. Specialized, formal and methodological research is easier to explain, easier to measure and far simpler to judge. It is therefore correspondingly easier to emphasize and easier to fund. Indeed, the reasoning which interprets science as dealing only with observables and with mathematical laws and principles derivable from observables, indicates that this is the type of science we should be funding.

The interpretation of science as a creative art suggests that we are wasting both our enthusiasm and our money. We should be funding the man with ideas. We should be supporting men who are attempting to understand the interrelationships between the disciplines and who are attempting to find unity in variety.

Possibly there are no such men to be found today, even if by a bureaucratic miracle we *were* willing to fund them. Yet men of this caliber always existed in the past. Historically, we know they have always existed, but certainly not always in the same society and not continuously. If conditions and orientations in current Western society make it impossible for such men to receive inspiration and support, then so much the worse for us and we should look to ourselves. One aspect is certain however: the empirical approach to the sciences, relying primarily upon collected data, is most important only in the infancy of a science.

The benefits to emerge from a re-assertion of an understanding of science as much more than experiment and method may best be illustrated by consideration of the social and life sciences. The so-called "social sciences" have long been handicapped by the belief in the power of "scientific method" and by the use of sterile models drawn from nineteenth century empirical science. The laws and principles derived from this approach have been limited in conception and temporary in applicability; they do not provide a recognizable understanding of human nature or activity. The trust in empirical science holds true for such related areas as chemotherapy and most environmental problems. There has not been a sufficient understanding to allow these disciplines to develop anything more than a superficial, trial-and-error, statistical approach. Perhaps a new approach to understanding will broaden the theoretical context of the "life sciences" by recognition of similarities in the laws of nature which pertain to all areas of science, physical and biological.

Despite the behaviorists who emphasize stimulus and response, and the Freudians who point to the subconscious, there does not seem to be any theoretical understanding of human or even animal behavior. Our knowledge consists of specific instances of observed behavior, with inductions from these observations to the so-called "laws of behavior." This knowledge is an early stage of investigation, where all of the difficulties of the problem of induction are running rampant, if sometimes unappreciated and neglected. In addition to the problem of relating the general law of behavior to the observed animal, the behavioralist approach cannot even touch fundamental questions such as what is human consciousness, what is mind, or how it is possible to draw different conclusions from the same set of stimuli.

It can be argued that these investigations are in their empirical infancy and we cannot entertain such philosophic questions as yet. But this argument is specious because such fundamental questions are currently proclaimed "outside the realm of science" and little indicates that the current formalism will be expanded to include these basic questions. A new approach will certainly not insure that the "ultimate answers" will be found, but at least some basic questions will be addressed once again in a serious manner.

Neither behavioralism nor quantum theory applied to biology seem to offer rewarding avenues of approach to a deeper understanding of the social and life sciences. These formalisms have and will continue to contribute tremendously to our detailed information about life. But they seem to set unsurmountable limitations on understanding by their insistence upon observables. The usual interpretation of quantum theory, of course, indicates that we can never know life processes with an exactness beyond the limits of the uncertainty principle, much less understand them.

A different approach to the social and life sciences might perhaps yield a new understanding which will serve both to re-order our current information and re-direct lines of research. Perhaps the information now available and the present clues before us are already sufficient to enable someone of powerful intuitive and creative ability to refine our understanding of the basic questions of life and behavior. We will never attain a complete understanding, nor, to use Kant's term, can we ever know the "thing in itself." We nevertheless must not negate the need for the search for increasingly broad understandings of nature.

There is no reason to believe that behavioralism is the last word in the life sciences and that the quantum theory offers a final understanding of the physical sciences. Historically, last words are seldom final. However, an end point in understanding may be reached by specific groups of people or civilizations. Because the West, with its current concerns with epistemology and behavioralist or mechanistic biology, rejects the notion of a deeper understanding of the "inner harmony of nature," it by no means follows that this understanding will not be forthcoming by men somewhere. In the past, destruction of life has been the only way of insuring the absence of new human creativity.

Since it has never been possible to isolate the exact direction from which new understanding emerges, some general indications of possible new approaches may not be deemed inappropriate.

Perhaps a new understanding will be aided by the revival of serious metaphysics—not the metaphysics displayed as an artifact or a curiosity within the context of epistemological concerns, but a metaphysics which attempts to re-assess the mind-body problem and to address the questions of the nature of matter and the nature of spirit and mind. Perhaps the study of the metaphysical implications of the relationship between matter and energy will yield some awareness of such current problems of the physical sciences as field conceptions and matter, and extremely high energies and small distances. Perhaps the problems of the nature of the nucleus, the energy it represents and the forces which hold its constituents in uneasy stability might fall into a coherent order if seen with a deeper appreciation of the nature of energy. Such problems are really ones of "being as being"; they are metaphysical problems, ontological problems. Historically, metaphysics has always been closely associated with natural philosophy in periods of deepest scientific creativity.

One may reason that no hard evidence indicates that philosophy will lead us anywhere, let alone in rewarding directions. The moribund nature of most modern philosophy is there for all to behold. A skeptic may argue that there are many well supported reasons to believe that our current approach to the physical world is correct and in fact the only one worth serious consideration. However, the modern rejection of metaphysics, based on our work with the limitations of knowledge, is based on an unsupportable position. Our modern rejection of metaphysics is based on an assumption which transcends the limitations on our knowledge itself, that assumption being that the limitations discovered and elaborated in epistemology are absolutely true. This assumption becomes our new and only truth.

Not only is this position to be regarded as overly hasty in the light of historical occurrences (the modification of nineteenth century confidence in classical science, for example), but it serves to reinforce the gloomy despair into which many, if not most, modern humanists were plunged by events in the first part of our century. Our current orientation denies the importance of transcendent values and beliefs, both in science and in all intellectual activities.

We seem now faced with the standard choice between agnosticism and belief. Should we allow ourselves to accept the limitations of our senses, or should we look to the re-emergence of a natural philosophy for both a broader context for the solution of our technological problems and for a broader understanding of man and man's place in nature? One choice involves science as a method and as a tremendously successful gatherer of information and a solver of problems; the other involves science as one of man's most creative arts.

NOTES

(1) J. Bronowski developed this line of reasoning in his work *Science and Human Values* (New York: Harper & Row, 1956).

(2) J. Bronowski, *Science and Human Values*, p. 20. See Bronowski's work for a full development of the concept of science as a search for unity, as well as for the presentation of Sir Karl Popper's fable.

(3) A. Einstein and L. Infeld, *The Evolution of Physics* (New York: Simon & Schuster, 1938), p. 294.

(4) Michael Polanyi, *Science, Faith and Society* (London: Oxford University Press, 1946). New edition (Chicago: Chicago University Press, 1964), p. 14. See Polanyi for the full development of this concept of intuition and science. See especially his work *Personal Knowledge*, 1958.

(5) Polanyi, *Science, Faith and Society*, p. 14.

(6) Einstein & Infeld, *The Evolution of Physics*, pp. 4-5.

(7) *Scientific American*, CCVIII (May 1963).

(8) G.H. Hardy, *A Mathematician's Apology* (Cambridge: Cambridge University Press, 1940).

(9) J. Bronowski provides this quotation and a full discussion in his *Science and Human Values*, Chapter 1, especially, pp. 16-20.

(10) A. Einstein & L. Infeld, *The Evolution of Physics*, p. 296.

(11) This passage appears in the Introduction to Erwin Schrödinger, *Science, Theory and Man* (New York: Dover, 1957). Intro. and ed. by James Murphy. Originally published by George Allen & Unwin, Ltd. in 1935 as *Science and the Human Temperament*.

(12) Erwin Schrödinger, *Science, Theory and Man*, pp. xvii-xviii.

COMMENTARY

Albert Crowell
Chairman, Department of Physics
University of Vermont

I certainly agree that science is not the same thing as technology, nor is it magic, although science has unfortunately acquired some aspects of both in the minds of many people. Technology however has not only drawn upon science to accomplish its ends, but technological considerations have also led to basic scientific discoveries. A classic example is Carnot's discovery of the second law of thermodynamics following his study of the efficiency of heat engines. As for science and magic, part of the modern rejection of science comes from a confusion of the two. People are disillusioned because the magicians of science have failed to deliver the desired miracles on demand. Yet the wonders they do deliver have a price tag beyond dollars and cents.

As Professor Steffens reminds us, neither progress nor miracles are the business of science. Science is, or at least should be, the search for an understanding of nature. It is an effort to acquire a perception and a description of the underlying unity and harmony in the multiplicity of natural phenomena. This effort is not, as is popularly thought, a formula known as the scientific method by which any computer could lead us from the known to the unknown. Nor is it the collection of meaningless data. Rather, science proceeds through the undefined and unconscious processes we associate with intuition, and its success is judged by the aesthetic qualities of beauty and harmony, and by its ability to bring order to the world of our impressions. Science, in short, is a creative art. At least I suppose that this description makes it so, although I am not sure Professor Steffens has defined creativity or any other kind of art, and I am certainly not going to attempt a definition.

Professor Steffens suggests that all is not well in the house of science, public misconceptions aside. Although science should not be concerned with the collection of aimless data, or with the solution of formal problems in the isolated disciplines of science, scientists themselves seem to be concerned with just these things, partly perhaps from the influence of federal grants and partly from such pressures as the "publish or perish" syndrome. Furthermore, he suspects that modern scientists have become trapped by faith in quantum mechanics much as their predecessors were caught by classic mechanics and electro-magnetic theory. He calls for a new approach, or perhaps a return to the old one, of natural philosophy.

If one visits the laboratories of many scientists and asks what they are doing, they may say that they are trying to synthesize this or that molecule, trying to unravel the reactions vital to the physiology of some animal or other, or are perhaps counting particles emitted by some obscure radioisotope. The published results often exhibit to the uninitiated all the inspired creativity of a page in a telephone directory.

There is a well known parable about three workmen who were asked what they were doing. The first man was chipping rock, the second was helping to construct a cathedral, and the third was building the House of God. As a member of the union myself, I would like to praise rock chippers. I do not know any scientist who collects numbers for the sake of collecting numbers. The seemingly pointless collection of data is a fundamental part of the search for basic unities of the universe. Certainly the conceptions and theories are intuitive acts of creativity and are the goals of science, but great theories are produced only by great philosopher-scientists, just as great cathedrals are designed only by great architects. Both great scientists and great architects are in short supply. Even when great creations of beauty and profundity are conceived, they are not cathedrals and they are not science, until the rock chippers have done their work. The scientific theory and the cathedral must not only be beautiful in design, but the theory must indeed help us to understand the universe and to erect the cathedral. It is the job of the ordinary scientist to discover whether or not a given concept helps us to comprehend the universe or only the mind of the theorist.

The role of the common scientist is not simply that of comparing nature with the image conceived by the giants of science. Professor Steffens has claimed that original creative science is natural philosophy and that it is not restricted by proscriptions of observability or non-observability. This claim is quite true. Yet the knowledge of what has been observed, or not observed, is an important contribution to the act of creation. Newton commented that he himself stood on the shoulders of giants. Today the giants stand on the shoulders of those of us of modest stature. We may give the impression of pointlessly chipping rock, or, like Newton in Blake's picture, merely playing with pebbles in the sand, but we too believe we are helping to build a great cathedral.

Are scientists confined by a dogmatic faith in quantum mechanics as the final word? Do they really believe that the unification of electro-magnetics, gravity and the principles of life must be sought within the context of the uncertainty principle? Quantum mechanics has its own unity and harmony, and is no more in opposition to understanding than were its predecessors. By encompassing Newtonian mechanics, it widens the ability of mechanics to describe the microscopic as well as the macroscopic world. I am confident that new understandings will be found, and indeed the lesson in the discovery of relativity and quantum theory is that change and greater generalization will come, and come at the expense of existing preconceptions. There is reason to believe, based on historical perspective, that freshness of approach and the emergence of deeper understanding will appear as cultural changes occur. Science does not blindly progress; rather it, like other creative arts, is a product of the culture of society. It is no accident that the generations which give us the abstract graphic arts also give us an abstract science. We will see not a re-emergence of a natural philosophy but a continuous emergence leading to a new, broader understanding of the universe and man's place in it. We must not be surprised if a more abstract world-view contains even greater puzzles than the uncertainty principle. Tomorrow's understanding will not be today's, and the universe will still be mysterious. At least I hope it will be, since the unknown is what keeps scientists at work.

We do need to support the men who are attempting to understand the interrelationships between the disciplines and who are attempting to find unity in variety. Such people do have the support and encouragement of much, if not most, of the scientific world. We are not faced with a divisive choice "between agnosticism and belief." Science is a gatherer of information and a solver of problems, and it is also a creative art; it has not ceased to be so, nor do I believe it will.

Whether or not the new philosophy will come from revolution or evolution, we should not forget that if it comes, it will represent even greater power than we have now. Perhaps it will indeed be the key to heaven. Richard Feynman, one of the greatest living physicists, has told of visiting a Buddhist temple in Honolulu. At the end of the tour, the guide concluded his remarks with the words that the key to the gates of heaven also opens the gates of hell. Surely, one of the responsibilities of philosophers, natural or unnatural, is to help society to know which gate is which.

COMMENTARY

William E. Banton
Director of Manufacturing
Western Electric Company

Professor Steffens stresses that Western man has erred in his concept of science, improperly misconstruing most technological accomplishments as science. Although I agree with him, I disagree that confusing technology with science is inappropriate. Knowledge for knowledge's sake is sterile. Applied knowledge or technology has created the world in which we live.

It is interesting that Professor Steffens places the time of this confusion during the Industrial Revolution in the middle of the nineteenth century. Another interesting parallel to this phenomenon is the evolution of the technical university in the Western world. In the universities at Bologna, Paris, and Oxford in the period from 1000 A.D. to 1700 A.D., a new kind of curriculum emerged from the training of the clergy. It was a two-pronged liberal arts program in which the physical sciences were taught concurrently with the arts and religion. Towards the end of the eighteenth century, physical science began to emerge as a subject not subordinate to the other curricula offered in universities.

In the United States, education at the university level followed the European pattern closely. After the founding of Harvard in 1636 and Dartmouth in 1770, nine liberal arts colleges were created which were closely patterned after Cambridge and Oxford. Courses in physical science were minimal or non-existent. Only the College and Academy of Philadelphia, which later became the University of Pennsylvania, included a considerable amount of mathematics and science in its curriculum.

The first engineering or technical curriculum was established in America at West Point in 1802, and the first university school of engineering was established at New York University in 1854. Technological or, as it is called today, "engineering" education began to evolve in the middle of the nineteenth century. This educational environment, coupled with the market for those educated in technological fields, fueled the Industrial Revolution which in turn created a demand for engineers or technologists and a greater integration of science and technology. Is this bad? We have the highest standard of living in the history of man and at the same time the highest levels of pollution ever known to man.

I would also like to examine what creativity is. What causes an Einstein, a Maxwell, or any other recognized genius of science? Is the act of creativity like the falling apple that hit Sir Isaac Newton, causing him to discover gravity? Is creativity an accident, or is it a more rational process? Jacob Bronowski in *Science and Human Values* said:

> In the act of creation, a man brings together two facets of reality and, by discovering a likeness between them, suddenly makes them one.

Jacob Rabinow in *The Process of Invention* quoted an article by William B. Shockley who said that creativity is not a simple functional relationship to the workings of the brain; if it was, it would indeed be more or less equal among most people. He postulated that, if an invention requires the combination of four ideas, and a man can put the four together, he may invent something. A man who can only put together three ideas will never invent. Shockley concluded that small differences in the *idea associating* abilities of individuals make a tremendous difference in their creativeness. Stated mathematically: creativity increases as the *log normal* distribution. Shockley's theory of idea association gives us a plausible theory of creativity and what causes an Einstein. In some colleges of technology, this theory of idea association is taught in courses entitled, "Creative Engineering" or "Value Engineering." How much "Creative Engineering" can be taught, and how much of it is related to a person's native ability? Can our modern team-research ever replace an individual act of creation? In industry we try to strike a reasonable balance between the individual creator working in his chosen discipline and the problem-oriented task forces directed toward rather specific goals.

In conclusion, a quotation of Harvey Brooks, Dean of Engineering and Applied Physics, seems appropriate. In his C.P. Snow lecture given at Ithaca in 1971, he stated:

> The discussion of science policy in the last three decades has too often confused necessary with sufficient conditions. A strong basic science is a necessary condition for a strong economy, a liveable environment and a tolerable society. But it is by no means a sufficient condition.

It is this test of sufficiency to which we are all addressing our current activities.

COMMENTARY

Robert P. Clagett
General Manager of Material Planning and Merchandise
Western Electric Company

Rather than examine the problems of creativity in Professor Steffen's paper, I want to examine two practical problems: 1. how the universities might structure their course content and approaches to educating the scientist, and 2. how funding the work of the natural global philosopher might be accomplished and where such a global thinker might be employed.

It is not at all clear to me what kind of education will create a future Einstein, Faraday, Kelvin, or Maxwell. In fact, it would appear that the greatness of intellect in such individuals would make it rather difficult to design an educational system that would prevent them from achieving significant contributions. Men of such intellect will, rather, take the formal education of which they can avail themselves and supplement it with their own energy and interest in any field that will add to their total store of information and philosophy. Such men could be helped by the manner of their formal education, but I rather object to the idea that we should structure our education to create, if we could, a small group of geniuses.

However, my support for formal education for all does not

lead me to conclude that Professor Steffen's suggestions are either impractical or irrelevant to today's world. In my own experience, quite often the experts in a given field were the ones who were only certain about what could *not* be done. I have had experts tell me that the reason I accomplished something was that I did not know I could not accomplish it. At other times I have found that reference to experts and their books, such as the *Handbook of Chemistry and Physics*, quite often told my associates and me that we were "on the wrong track" and that the job to be accomplished could not be done in the way we were proceeding. In those cases it took us a long time and it required much effort to get back where we again questioned expertise and found that it was incorrect in one way or another. We then went ahead to complete our task. My experiences lead me to think that some of our scientific training has become so compartmentalized, so narrow in problem-solving that we indeed *do* learn only what cannot be done. We should open our courses to a broader view, for current knowledge is only what has been learned to date and is quite likely not to be complete and final. The kind of instruction that allows for a variety of creative solutions is the kind that will allow a questioning and searching attitude to be developed in a student.

To suggest that our courses should somehow imbue students with an appreciation for the metaphysical and even with the use of the metaphysical in scientific research seems to me overly ambitious. I certainly applaud studying the natural sciences and the arts as well as taking courses on man's ability to live with his fellow man and man's ability to live with the natural world. This liberal study would enhance any scientific education. However, the current interests of students lead me to believe that in the absence of that kind of study, thinking young men and women will still avail themselves of that information and will question the narrowness of the teaching of some of our disciplines. However, their education would be enhanced if a university offered such courses as part of its curriculum.

As to where such study and research might be sponsored, where can a person get away from the everyday problem-solving to think about the interactions of disciplines and the larger questions that confront us? The university and university-associated programs are about the only means by which we can expect such work to continue. The Bell System in its research organization, the Bell Telephone Laboratories, does indeed fund and sponsor what might be called basic research. However, that basic research is concerned with communications theory and materials investigation. Those things, although basic and not *directly* applicable to current problems of telephony, do indeed find sponsorship because they are allied to that field. No business can really sponsor much of the kind of investigation which Professor Steffens outlines, because of the competitive nature of business. If a firm were to sponsor and fund a great deal of investigative research, which I think we agree would not have an immediate or even a long range *direct* benefit to that firm's endeavors, the sponsoring business would be carrying a financial burden that would be greater than the investments of another firm without that burden. Consequently, the sponsoring firm's product would become overpriced and non-competitive. The sponsor would sooner or later go out of business. The loss of a business does not help the industry, the nation, or our society. Rather, it is the role of all business to support universities with grants-in-aid, with scholarships, and other assistance that allow schools to select and sponsor people and investigations.

The other possible source of educational support is of course the government, but I view governmental sponsorship of education in much the same way as I do assistance from business. I would be concerned with the government's selection of projects to fund. Pressure of the constituency on the Congressmen themselves point to such funded projects as impractical and as a waste of taxpayers' money. Government should help to support universities at whatever level the nation can accommodate itself, but that level should not involve outright direction of projects.

In conclusion, a change in direction is needed both in university education and the way in which people in the sciences look at their calling. We have accomplished much in a material way, but little in our ability to live with ourselves and the natural world. It is one thing to say that we should shift the emphasis of our education and quite another to then decide how to create the opportunity to pursue the result of that education. Where do we allow the global thinker to function? Where do we fund such endeavor? The only practical place is within the universities, since business must continue to strive to be profitable in order to have funds that it may use to support university education. Similarly, government is not the appropriate base for such sponsorship, because of its predilection to direct projects and because of the pressures to restrict spending in areas with little short-term output.

RESPONSE

Henry John Steffens

What interests me most about the commentaries of Mr.
Banton and Mr. Clagett are their attitudes toward science.
Their interpretation of science is clearly quite different from
my own.

Mr. Banton's statement that knowledge for knowledge's
sake is sterile seems to strike at the heart of what I consider
to be our disagreement. Knowledge for knowledge's sake may
seem sterile when compared to the standards of the production of material goods, the amelioration of societal problems,
or when compared to feeding or clothing people. However, a
major point of my paper was that without the general
knowledge which must precede practical application, we are
locked into the limited current mode of formulating solutions to our problems. We certainly have the option to
continue using the same kind of technology—using it and
reusing it. Historically, however, changes in the way of doing
things in society have always come from changes in the way
of knowing things in that society. We may agree that we
should change our way of doing, but to do that, we must

change our way of knowing. Far from being sterile, knowledge is the essential element in the whole process of genuine change.

Mr. Clagett's statement that relating metaphysics to scientific research "seems overly ambitious" is one I want to discuss further. One need not do metaphysics in order to be extremely creative in designing new kinds of electronic circuits, in finding new ways to use transistors, or in developing new kinds of computers. I do not mean to imply that there is any difference in the kind of creativity that one brings to the design of a new piece of equipment or to the creation of a poem or a scientific theorem. However, in periods of deepest scientific creativity, where the West succeeded in fundamentally changing its orientation and stance toward the world, metaphysics was always integrally related to science or natural philosophy. I am willing to concede that metaphysics is perhaps of little aid to a man with a specific problem to solve. Yet, fundamental changes are dependent upon the depth of our understanding of the natural world, and that understanding must be both in terms of metaphysics and in terms of epistomology.

My understanding of the relationship between understanding and change did not lead me to a misunderstanding of the function of Professor Crowell's rock chipper. The rock chipper is certainly of vital importance to a whole project. Notice, however, the point at which the rock chipper begins to chip his rocks; it is after the mountain has been discovered. It is after the philosophy has been uncovered and revealed. The image is very suitable and might be extended a bit: had the rock chippers not received the block of granite from the mountain, they would have had nothing to do. Fundamental creativity and discovery involve intuition and imagination in an extra-observable, extra-methodological way. Once this kind of creativity appears, then there are many things for many people to do on all levels of creativity.

Finally, in the nurturing of creative imagination, we have to be careful not to submerge the creative individual in a society which has focused entirely upon behavioral and technological objectives. A genius is a genius, and it is very difficult to surpress genius, although of course it can be done. Historically, geniuses have directed their genius in a variety of directions. For example, French science at the end of the eighteenth century, just before the French Revolution, was a theoretical science par excellence. French theoreticians were among the finest in the world; French mathematics was really superb. Then the French Revolution, the European wars, and Napoleon's rise to power occurred. Napoleon had an excellent army, but it was dependent on the accurate use of artillery. The French educational system was involved in an interesting way with the accuracy of the French artillery. The system was structured so that the so-called secondary schools, or *lycées*, had two basic kinds of curricula: the belle lettres curriculum and the mathematics curriculum. If a young Frenchman was inclined toward science, he took mathematics and aspired to the *École Polytechnique*. During the course of some twenty-five years of war, the Minister of War decided that because so many young artillery officers were being killed, he should induct his artillery officers not only from the *École Polytechnique* but also from the *lycées*. Finding men in the mathematics curricula, he commissioned them in the army and sent them to a glorious but short career in battle. It did not require genius to perceive—although many geniuses did perceive—that one should not opt for the mathematics curriculum in the *lycée*. It is no mere coincidence that during the generation after Napoleon, the French theoretical sciences virtually died. No new generation of physical scientists emerged to compare with the generations before the wars. However, belles-lettres suddenly rebloomed with political criticism, poetry, and drama. It does seem that genius will out, but the directions it takes can be heavily influenced by the societal conditions of the time.

The question of what can we do with the creative, imaginative individual has been raised. Mr. Clagett suggests that we should not fund him through government but should place him in a university, because business cannot support him and remain competitive. Again historically, Galileo was sponsored and funded by government. He flourished because rich princes supported him. After he assembled the telescope in 1609, he was pensioned by the Venetian government which was very interested in the telescope. The Venetian officials were not interested in Galileo's discovery that the moon has valleys and mountains and that Venus had phases. To the government, the telescope was important because it enabled men to see a ship two hours earlier than the naked eye could see it. Whether it is found in Galileo or a nuclear scientist, genius can, in fact, be supported with governmental aid, if that aid is handled properly.

Secondly, universities can fund genius, if arrangements are managed appropriately. In the late nineteenth century, the Universities of Glasgow, Oxford, Edinburgh, and Cambridge all supported men on their university staffs by simply allowing them to do whatever they pleased. It is, unhappily, highly unlikely in our current societal condition that we will fund a creative genius properly. If creativity is important in changing the directions of both science and society, then we had better look more closely at our societal and educational attitudes toward genius and our support of it.

I want to conclude my commentary by returning to knowledge for knowledge's sake to make a personal statement about the art of being an historian and its contribution to knowledge for knowledge's sake. Properly done, history must be an art and must present an interpretation of the past as seen through the eyes of the historian. That interpretation goes beyond a so-called "objective" statement of "the facts as they actually happened." It is true that if history is an art, then history should produce an interpretation of the past for its own sake. What society does with this interpretation is really the crux of the matter. If a society ignores its past—and very firm indications exist that modern society in the last two decades has been profoundly ahistorical—that society is literally lost. If knowledge of the past is ignored for its own sake, then people cast themselves completely adrift in the world. Historical knowledge offers the possibility of understanding a present stand, of understanding where people have come from and where they may be headed. Historical knowledge enables us to develop an intelligent perspective, which can provide both firm foundations and guidance for action. This historical knowledge is best when it approaches a situation of knowledge for knowledge's sake, because then it is least distorted, propagandistic, and misguiding.

Historians very clearly have biases. The historian is clearly subjective when he writes history, but that subjectivity is an essential ingredient in historical work, just as much as metaphysics is an essential ingredient of scientific creativity. People who say that they are doing "objective" history are either fooling themselves or attempting to fool the public. There is no such thing as objective history, for a variety of reasons. There is, to be sure, a range of historical facts that indicate that something has happened. These kinds of facts are the most simplistic and useless kind of information that we can possibly amass. After all, who cares exactly when so-and-so died or so-and-so was born, unless one attaches to that fact some reason for caring for it. If there is a reason for caring, then there is usually subjectivity. Facts are available, lying around like stones in a field, but they do not mean very much until an historian has some reason for looking at them. Even at the level of the initial use of facts, subjectivity enters into history in a fundamental way. The level of seriousness in history is a direct function of how well one can perceive one's subjectivity in looking at the so-called historical facts. For this reason, we can place a Marxist historian next to an historian of art, next to a political historian, next to a social historian—all writing about the Renaissance, for example— and draw a useful interpretation of the past from all of them by observing how their various accounts cohere. All accounts will be in some manner subjective, but because we can recognize the kinds of bias that each has brought to the study, their accounts have usefulness. We need to be concerned not so much whether history is objective, but in what ways it is subjective. If we abandon the naive assumption of absolute objective truth, we shall be able to see the richness and many-faceted aspects of knowledge of the past.

Objectivity does enter into science when the scientist attempts to relate the kinds of theories that he formulates to the world of sense experience. One of the great lessons of the twentieth century is that science can be objective in that it can tell us whether our theories about the world are wrong. Yet, it cannot be considered completely objective. We can never tell whether our theories are absolutely right. The issues raised here are related to the problem of falsification and confirmation. Being right would imply that we have really reached the truth. All of the lessons in the change from classical to modern science indicate that the "whole truth" about nature is illusive.

BIBLIOGRAPHY

Bronowski, Jacob, *Science and Human Values* (Harper Torchbook TB 505G, 1959).

Bronowski, Jacob, *The Common Sense of Science* (Vintage Books V-168).

Conant, James B., *Modern Science and Modern Man* (Doubleday Anchor Books 10, 1953).

Holton, Gerald (ed.), *Science and Culture* (Beacon Paperback, BP 250, 1967).

Holton, Gerald, *Thematic Origins of Scientific Thought: Kepler to Einstein* (Harvard University Press, 1973).

Polanyi, Michael, *Science, Faith and Society* (Phoenix Book P155, 1964).

Schrödinger, Erwin, Science, *Theory and Man* (Dover T428, 1957).

Wagar, W. Warren (ed.), *Science, Faith and Man* (Harper Torchbook TB 1362, 1968).

SCIENCE
AND
SOCIAL
RESPONSIBILITY

INTRODUCTION

Despite the historical and philosophical dimensions of the first three sections of this book, the authors found a common focus on the critical problems of contemporary civilization, especially those problems seen as the by-product of technology. They argued that a clear understanding of the creative, intuitive nature of science as distinguished from the problem-solving approach of technology is crucial to coping with current world problems. Yet when in discussing the meaning of science in relation to social responsibility, Donald K. Conover and the three commentators on his paper dispense with these distinctions. When the world's immediate problems and fears are considered, the issue of power is thought to be more crucial than philosophy.

Professor Felt had asked earlier in this volume, which set of priorities would be imposed on current and future technological capability? Yet, with clear calls for change on all sides, no one had focused on the method of selecting the priorities other than keeping current ones or on gradually changing through the ferment of widespread communication between disciplines or groups within society.

Mr. Conover restates the prevalent view that we are living in a time when we fear that misdirected technology coupled to a faulty social structure has grown to a proportion that represents a threat to our comfort and possibly our survival. In this context of immediacy and apprehension, the fine, precise philosophical distinctions between natural philosophy and scientific technology are ignored. "What we call science," according to Mr. Conover, "fits various definitions which depend upon the frame of reference from which we approach them." He holds that science and social responsibility are related in an end-and-means relationship.

John W. Engroff is more direct when he notes that the time between pure research and its application has been so shortened that the distinction between them has become obscured. "For our purposes today . . . we can combine pure science and applied science and simply call the combination, 'science and/or technology.' " Such a point of view implies a despair that solutions to current problems may not be found in the art of scientific creation and the development of a philosophic construct.

In disposing of the distinction between science and technology, Mr. Conover argues that science provides the means by which social responsibility determines the ends. Consequently, science becomes valueless as well as agnostic. His position is very different from that presented by Professors Williams and Steffens, who view science as the product of personally accepted value systems.

To help find answers for critical world problems, Mr. Conover urges appreciation of technological advances, especially those in "various methodologies in systems analysis (which) are being employed to give us a more accurate picture of what is already taking place and where present trends might take us." He adds that the systems analysis "provides a hypothesis for understanding cause-and-effect relations."

Having such a potentially capable tool as systems is not

enough, however, because of man's rigidity and general "unwillingness to confront realistically the need for social responsibility." This failure to confront social responsibility is exacerbated by "the language of system analysis, the mathematics it employs, and the heavy role of computers (which) all conspire to erect cultural barriers between the systems technologists and the rest of us."

Having developed, as George V. Cook did, the crucial importance of systems analysis, Mr. Conover does not agree that such analysis will be used to bring the two or three cultures together in a progressive harmony to attack social problems. Unlike Mr. Cook, Mr. Conover believes that the systems analysis approach may ultimately only serve to produce a further anti-technical reaction.

That reaction and its attendant by-products may be the greatest difficulty in our addressing ourselves to obvious current problems. Mr. Conover presents a gloomy picture of the division in basic attitudes between individually and organizationally oriented people and the resulting "contradiction in personal orientation about the relation between science and social responsibility." This polarization "divides our thinking and is the major block to an integrated resolution of problems. Moreover, to the extent that the division has become a formal part of society, isolating the seekers of truth from the doers and builders, one may wonder that we have done as well as we have."

To bridge this division which both he and Mr. Cook see as crucial, Mr. Conover suggests that men must face the crucial problem of spirit. Our current formulae are wrong, "because they are incomplete, and we have denied the validity of our spirit as part of the equation." In dealing with the spirit, Mr. Conover comes close to agreement with Professor Steffens' warning that hope lies in a new philosophy. Yet to Mr. Conover, metaphysics must be rejected as an approach which is too slow for a quickly changing world.

The three commentators, each in their own way, accept Mr. Conover's premise that the proper application of technology, infused with the necessary spirit, offers the clearest hope of meeting human needs. They concentrate on the values that would direct the application of technology. John Engroff focuses on "the use and abuse of power and . . . the fact that neither science nor technology is politically neutral or value-free." Unlike Professor Steffens, who postulates that science is not only the product of values but also is outside the context of judgments of right and wrong, Professor Engroff finds science interwoven with the economic, social, and political fabric of civilization. Therefore, science is "intimately linked with power." Professor Engroff wonders how the use of science is safeguarded. If science and technology are controlled by corporate capitalism, how can a runaway technology be reformed without overhauling the economic system which depends upon it?

Willard M. Miller is in close agreement with Professor Engroff. He too argues that although scientists want to claim the comfort of value-free orientation, their science is closely related to the values of their society. "The only socially responsible context for scientific activity," Professor Miller asserts, "occurs when it relieves suffering and betters the quality of life." He finds our present scientific efforts socially irresponsible, for in the American socio-economic system, consumerism, imperialism, and the inability to change in large part are conditions closely related to the misuse of technology.

Professor Miller calls for revolutionary development of a genuine democracy which can reverse current priorities in a period of non-growth. That genuine democracy, which would provide the only proper basis of assigning the priorities by which technology is applied and civilization is guided, is similar to Mr. Cook and Mr. Conover's idea that the full participation of all relevant factors come into play in a genuine democracy. How a genuine democracy would come to conclusions and establish priorities is not made clear.

Stanley M. Grubin endorses the necessity to consult a wide variety of sources and people in setting priorities for our technological society. "Man is becoming," he notes, "less and less likely to sit back and allow the literary scholar or the scientist to establish society's goals." The entire problem of making decisions about priorities is further complicated by the speed of change. Unless "solutions to problems can be developed and implemented almost instantaneously, the definition of a problem will erode during the problem-solving stage."

SCIENCE
AND
SOCIAL
RESPONSIBILITY

D. K. Conover
Director of Corporate Planning
Western Electric Company

There are many ways to approach science and social responsbility. Since most of them lead to arguments about science *versus* social responsibility, their only value is to bring one's particular prejudices out in the open where they can be picked apart. As sport or therapy, lobbing rhetoric back and forth may serve a purpose, but it is unlikely to yield much in the way of new understanding.

Another possible way is the dispassionate approach. One lists all the pros and cons for each side, gathers every conceivable historical reference, and presents a scholastic treatise that allows each person to draw his own conclusions. I am unqualified to take either approach. I am not sure there is an historical precedent for the kind of problems involved in the current concern with science and social responsibility, and I find it impossible to be dispassionate about issues that deal with survival, especially my own.

We are confronted with a nice situation. According to some, a good case can be made for the idea that runaway technology is going to make our planet largely uninhabitable.

Others have a cure for an uninhabitable environment, but they want to make the earth a place where most of us would not want to live. I find no comfort in trying to choose between these alternatives.

I do not have a choice, but I do have a viewpoint that suggests a choice is possible. By addressing it, perhaps we may be persuaded that our concern over symptoms has led us to neglect looking at causes. If we can get a better understanding of why the relationship of science and social responsibility represents a problem, we will be much closer to finding an acceptable way to deal with it.

To state the problem: we are confronting a situation in which we fear that misdirected technology coupled to a faulty social structure has grown to a proportion representing a threat to our comfort and possibly our survival. Thus, when we discuss science and social responsibility, we are really talking about their contribution to this threat or their potential for reducing it. Moreover, we are talking about their effect upon one another.

My thoughts about science and social responsibility will deal with their impact on living in a way that we might agree is desirable. This kind of desirability will involve enlarging our concept of social responsibility to include more than just providing for our material needs.

Professor Williams established the thought that while most science involves practical problem-solving, a tiny fraction of scientific effort, by its creativity and cosmic scope, changes our view of the world. George Cook considered the two cultures and developed the judgment that whatever differences specialization might tend to promote, there is a correlative duty to build communication bridges to promote and preserve understanding. And, Henry Steffens elaborated upon the unique and unpredictable character of real creativity which yields new philosophies and which applies to science as well as art. Each refined our understanding of what we mean when we talk about "science," and each had something important to say about our relation as people to a world in which the scientific perspective is but one of several viewpoints.

It becomes apparent that what we call science fits various definitions which each depend upon our individual frame of reference. Our definitions will vary as we discuss social responsibility. For example, the *Wall Street Journal* recently quoted Robert Meridian, then Assistant Attorney General, who, in discussing "revolutionary terrorists," said, "We're faced with an unprecedented problem—we're getting more people in government who feel they should be ruled by a sense of conscience." As it relates to social responsibility, Mr. Meridian's comment exposes him to some very different interpretations about what our government's problems really are.

Concern about definitions is legitimate and, because it is so knotty, makes the problem difficult to discuss without misinterpretation. I am not going to try to add definitions of science or social responsibility that are irrefutable. Rather, I am going to suggest that we view the two terms as connected in an end-and-means relationship. Although this view may only be one way of looking at the relationship, it is revealing in terms of understanding something about what is happening in our world as a whole and what approaches we might examine to begin solving some of our more serious problems.

Most commonly, we tend to think of science as a means to an end. As a simple-minded example, consider the hammer. It is a device employing mass, kinetic energy, and leverage, and it clearly represents a legitimate, though primitive, example of scientific creativity. What man did with the hammer, whether he used it as a tool to build things or as a weapon to bash in someone's head, is a question of social responsibility. Science provided the means, and social responsibility determines the ends.

The hammer is not a totally satisfying analogy unless we embroider our consideration of it with a bit of imagination. Suppose we could apply some of today's concerns about science to the question of whether or not this thing called a hammer should have been developed in the first place.

Recognizing that some ingenious caveman is fooling around with a new device, we might argue that he should desist because man is bloodthirsty enough without hammers, and who knows where things might end if everybody were allowed to wander around with such an advanced skull cracker. Although it would be handy to have for hunting and building, continued hammer development would be like opening Pandora's box; no one knows what terrors might come out of it.

Beyond what is known about socially responsible uses for science, a fear remains of what is unknown about the impact of changes that follow each new scientific development and about our ability to adapt and survive in the new world they introduce to us. How should we be looking at science as it contributes to the ultimate social responsibility, which is survival in a style hospitable to our humanity. If we deal with what is known, the means-and-ends relationship between science and social responsibility is a workable strategy. If we deal with what is not known, the issue of this kind of relationship degenerates into a question of our attitude about change and whether we are inclined to be optimistic or pessimistic.

Unfortunately, already enough is known to worry about without our inventing imaginary nightmares. Science has already provided technologies that may carry an ultimate impact that could be disastrous. Without any more discoveries and new inventions, what we already have, if improperly used, is sufficient to bring our species to its knees, if not destroy us altogether.

Almost daily, someone announces a forecast that provides a new scenario of doom. Some forecasts are little more than emotional outcries against change; however, some are not. Increasingly, various methodologies in systems analysis are being employed to give us a more accurate picture of what is already taking place and where present trends may take us. Because they have been effective in educating us to a better understanding of interactions between societies and their technology, certain realities are now accepted at face value.

For example, Buckminster Fuller's concept of "spaceship earth" highlights the fact that we are, for all practical purposes, contained in a finite world possessing finite resources. Because population and industrial growth result in increasing withdrawals from the resource supply, it is not a fantasy to worry about running out of resources.

A more complex forecast involving much of the same basic information is embodied in the Club of Rome's report, *The Limits of Growth* by Dennis Meadows and a team from M.I.T. This study simulates the interaction of social, economic, and technical variables in terms of level and rate relationships acting over time. A variety of different assumptions all lead to the same conclusion: continued population growth and capital growth will ultimately lead to a crisis in which our planet can no longer support the people trying to live on it. What the different assumptions influence most is the timing, not the nature of the crisis.

As interesting and important as they are, it is not my intent to dwell on the details of these forecasts. Instead I would like to examine the forecasts as scientific events which are connected with questions about social responsibility.

At the outset, we must realize that systems analysis on a global scale is a "soft" science. Depending on a combination of assumptions and relationships inferred, but not proven, an approximation of reality is synthesized and then operated as a model of reality. The results tell us what would happen in reality if the assumptions and relationships in the model were correct. We have no sure way of knowing how good any model is except by waiting until time has passed and the forecast can be compared with what really happens. The next best thing we can do is make some judgments about the model's reasonableness and do some testing to determine the sensitivity of the model to variations in the assumptions.

As a science, analysis has changed our way of looking at the world. Even if we disagree with a particular model, most of us will acknowledge the basic viewpoint of the world as a system in which interaction rather than independence is a key to understanding. Unfortunately, the language of systems analysis, its mathematics, and the heavy role of computers all conspire to erect cultural barriers between the systems technologists and the rest of us. These barriers affect what we do about this relatively new science and, as such, relate to the question of social responsibility.

In social terms, systems analysis (particularly, global fore-casts) provide two things. They give us an early warning that something may happen, and they provide a hypothesis for understanding cause-and-effect relations. Now let's consider the behavior that results.

One of the most popular responses is to zero-in on one variable and develop a simplified solution that presumes the other factors are benign. In my view, this is what the anti-technology movement is doing, for they would have us believe that by arresting new technical developments, we could avoid the problems being forecast.

This viewpoint implies that man is incapable of the self-discipline to regulate and use technology wisely. Obvious-ly, technology has been used foolishly, even criminally in some cases, but, can anyone seriously hold that misuse of technology is man's only field of indiscretion? Has, for example, our progress in medicine and nutrition been ir-responsible because it contributed to the population ex-plosion? I could just as easily state, and be on solid statistical ground, that more technology will reduce birthrate by fur-ther complicating our life and distracting us from an activity that is still pretty non-technical, the making of babies. Instead of sending food or medicine to overpopulated coun-tries of the world, maybe we should send television sets for late night viewing.

Another particularly popular response with the idealisti-cally inclined views of the overall picture presented by systems forecasters takes note of the fact that too much growth, unevenly divided, is a problem and immediately concludes that no growth and equal distribution must be the solution.

As typically presented, the no-growth argument either assumes that it refers to other people without substantially reducing our American standard of living, or it blithely assumes that no-growth and equal distribution will only reduce our luxuries (which is sometimes thought to be a virtuous goal all by itself). The first assumption depends on an unbelievable degree of stoicism by the poor of the world, and the latter fails to comprehend the real extent of inequality between different standards of living.

To approach equality in distribution with no growth would probably involve giving up such luxuries as flush toilets, central heating, refrigerators, preventive medicine, most secondary and higher education, most broadcasting and telephone communication, a substantial portion of our homes, as well as the more obvious things like private automobiles, recreational equipment, and most of our clothes. That list of items may be startling, but it must be remembered that it is based on the world average personal income of $465.00 a year.

At the other end of the spectrum of responses to gloomy forecasts, we have a set of ideas that tend to be popular among that group called the "Establishment." Recognizing that any real solution is likely to upset the status quo, a member of the "Establishment" usually either attacks the forecast as though it were the problem or simply closes his mind and hopes the whole matter "blows over." Both responses will probably be used against the Meadows study. It contains enough assumptions to keep nit-pickers busy arguing about how the study was designed so that sooner or later we will get tired of hearing the arguments and will be glad to bury the study without ever dealing with what we should do in the event its assumptions are correct.

Another popular response among people who are basically happy with life as it is in an exhortation to "try harder." There are two variations of this response: one non-technical and the other technical.

The non-technical strategy assumes a faith in people's ability to muddle through life. To improve the odds, it is usually suggested that we need to discipline ourselves to work a little harder and not to expect "pie-in-the-sky." Inasmuch as people are pretty good at muddling through wars and natural disasters, we probably should be more realistic, but such complacency denies that any problem can ever be big enough to make us question the need for serious change from "business-as-usual."

The technical response is the antithesis of the anti-technology one. It is summarized in the statement that "more technology is the best way to solve technology's problems." That slogan is a limited truth. It does seem obvious that the problems of an expanding, sometimes faulty technology are most likely to be solved through more and better technology. Moreover, because we are part of a society which has concentrated on technical skills, technology is an area of strength where a legitimate basis exists to look for dramatic and substantial improvement.

If we look at these different responses collectively, several things are apparent. First, our rigidity and unwillingness to confront realistically the need for social responsibility is an area of acute threat. If we persist in making science a scapegoat for all that is not right with the world, we are excusing ourselves from the responsibility of setting positive goals for what is right. Those who rely on technology as a substitute for needed change in social areas and those who are anti-technology are part of that group of people who are avoiding responsibility. Second, each response implies a judgment about underlying motivations of various people, usually those in positions of power. Each response also assumes a utilitarian role for science that can be used in ways that people judge to be good or bad.

Most often, the "bad guys" are seen as rich and powerful. They are perceived to use science for their own selfish ends or to ignore it when its findings point to a change in the status quo. Such a generalization is unfair, for it is often simply untrue in any sense. What probably approximates the truth is that applied science is simply running ahead of our willingness to effect the appropriate social change to cope with it.

Concerning the issue of science as a means to achieve social responsibility, what has changed to make this relationship a part of the problem is the acceleration of change itself. As a means, science must be guided by a philosophy. Moreover, that philosophy must permeate the society sufficiently to provide an overall sense of direction. When our abilities or our tools are more developed than our knowledge of what to do with them or why we should use them at all, problems are bound to occur, and they do.

Professor Zbigniew Brzezinski of Columbia University placed the issue in perspective when he said:

> Without science, modern philosophy cannot possibly supply answers to such concrete problems as ecology, survival, pollution, nutrition, even peace. Without philosophy, science would be directionless, possibly destructive . . . We have to ask why—that is the philosophical part. We also have to ask how—that is the scientific part.

Considering the need of science for philosophy, it is not so unreasonable after all that our thoughts about science and social responsibility are so often characterized by half-truths. The real task, as George Cook has explained, is for us to find ways of bringing together the values of many different human resources, whose specializations often impose particularized viewpoints. Only together can we seek the whole truth.

To conclude simply that science is good because it plays a part in the solution of our social problems, or to claim that these problems are simply due to an accelerated rate of change is not very satisfying. Such statements do not explain why we so often misapply science or avoid making socially responsive changes. It is not enough to identify the potential for solutions that have been available all along; we must also discover what is keeping that potential from being realized.

To do this, we must broaden our thinking. Professor Steffens appealed for renewed attention to metaphysics. Without belittling the long-term value of his approach, I think we can ill afford to content ourselves with a contemplation of combustion when our house is burning.

The simple truth is that we must behave in a better way than we do; we must develop an increased respect for the rights of man in a world grown more complex through the developments of man. This truth is not a scientific approach to our problems even though science may help us to understand them. It is more concerned with the spirit. We need to confront ourselves, because metaphysics has failed us.

Our scientific talent, whether pure or applied, may mislead us. It is not enough to be alert to the threat of a runaway technology. It is not enough to look at simplistic social goals as though people were just a herd of animals to be fed and cared for in order to keep their bodies fit and their behavior in order. Our concepts of social responsiblity must make allowances for the cultivation and preservation of individual taste. To do so requires more humility and honesty about ourselves than we have displayed in the past.

A few years ago, I attended a meeting on government planning in Paris. It was summer, and about fifty of us were seated around a large table in an elegant room belonging to an age long past. Floor-to-ceiling doors to the outside were opened for ventilation. Although the room faced a small courtyard, the sounds of the city streets drowned out most of what the speaker said. As we were leaving, someone asked why we had not met in an air-conditioned building so that people could hear what was being said. With a typically French gesture, the speaker acknowledged the practicality of the idea but said the move would have meant giving up the beauty of the room we had just left. Although the man was a close advisor to President de Gaulle, he added that the pleasure of being in that room was probably more important than anything he had to say.

That kind of pleasure is an aspect of living in which neither science nor social responsibility can provide answers. It involves our sense of being alive and being moved by that awareness. For some, science provides it, and when it does, we may speak legitimately about beauty in science. Beauty is the word we use to describe experiences that make us aware of truths which make being alive meaningful. Social responsibility permits our finding beauty by giving us the freedom to discover it.

Thus, we can see the contradiction in talking about science and social responsibility. If we are discussing society, science is a means capable of contributing to socially responsible ends. If we are discussing people, social responsibility describes the environment which allows the pursuit of science or any other form of truth which brings beauty to one's life.

The contradiction about the relation between science and social responsibility as ends-and-means provides a basis for understanding some of our difficulty in arriving at agreements on goals for the future. It clarifies a basic difference that has developed between the outlook of people whose orientation is geared to organizational achievement and those more interested in a personal quest for beauty or truth.

For example, it becomes logical to consider pure scientists, artists, and scholars as a group with a similar orientation. Their concern and satisfaction with the search for truth represents a goal or end, which can be called "beauty" because, regardless of the field, the awareness of beauty describes the end result of a successful effort. For this group, social responsibility means assuring an environment where the search for beauty can proceed with a minimum of hindrance.

Among engineers, business managers, and politicians, we can identify a group with a goal which is inherently social and organizational. If there is beauty in their goal, it involves the kind of truth revealed by people living in harmony. For this group, science, the arts, and scholarship are viewed as utilitarian assets to achieve the kind of harmony its members perceive as "social responsibility."

Differences in orientation explain, for instance, why business managers are typically horrified at the way universities are sometimes run. To them, the goal is harmony and efficiency in producing educated graduates. To scholars within the university, such harmony is irrelevant unless it provides the kind of environment where the search for truth, as an end in itself, is unfettered. On the other hand, a dean or university president who comes from a background in which a scholarly orientation was appropriate may face profound frustration working in a new job situation where a new orientation places him in conflict with his prior values.

The recently-graduated student who has adapted to the truth-is-beauty orientation in school may also suddenly find himself employed by an organization that asks him to concentrate on results which fulfill the *organization's responsibility*. Criticized for being too "idealistic," he may feel he is prostituting himself by changing his orientation.

The contradiction in personal orientation about the relation between science and social responsibility involves a polarization which divides our thinking and is the major block to an integrated resolution of problems. Moreover, due to the extent of the division between seekers of truth and doers or builders, one may wonder that we have done as well as we have.

It is not enough to increase our efforts to develop new philosophies, if the philosophers are so far removed from the rest of the world that their ideas never reach popular understanding or practical application. On the other hand, it is not enough to have honorable intentions in wielding social power in business or government, if we cannot envision more than a safe and orderly society.

The realities of power suggest that the greater danger from abuse lies with those oriented toward organizational goals.

For example, I was living in Chicago during the riots at the 1968 Democratic Convention. Although I do not defend Mayor Richard Daley and his role in the troubles of that week, I would like to suggest that the Mayor was acting according to a sincere sense of social responsibility as he interpreted the events taking place. When confronted by a situation that seemed to threaten the dignity and harmony of the community, his political party, and his administration—all the organizations to which he is dedicated—his actions were geared to protect those organizations. His actions involved a serious compromise of other values, but to him the integrity of the organization had become more important than the goals of the organization. Social responsibility was reduced to upholding the status quo at any cost.

In a less dramatic way, the status quo is maintained by people more concerned with bureaucratic procedure than in solving a real problem. Some of these people are in business organizations, and I would be surprised if they did not also exist somewhere on campuses.

Their problem is not a lack of social responsibility so much as a narrow interpretation of what social responsibility means. If social responsibility is going to stand for anything more than preserving our organizations, we must accept the reality that in the absence of inputs establishing principles for individual rights, duty to an organization will dominate. Thus, we agonize over our human history in Auschwitz, Dachau, Budapest, My Lai, and Bangladesh. If we are going to avoid these tragedies of history, our concept of social responsibility must find a better balance between the need for stability and the need for freedom. That balance will not occur without a coming together of leaders representing both organizations and ideas to search for a truth that is both visionary and practical.

Not long ago I sat on the banks of the Tagus River, looking across the water at the city of Toledo and seeing it much as it must have appeared to El Greco. Walking the narrow streets of Toledo or standing in the dusty majesty of the great cathedral, one forgets what time it is. Yet one is alive and aware of a fresh understanding of what living means.

However, by any practical test, there is no reason for Toledo to exist. The meaning and pleasure of knowing the city does exist is too personal to fit any of our formulae or programs for social responsibility. Toledo does not feed anyone; it produces nothing; it is absurdly inefficient; it is unnecessary. Yet it can make one's heart sing to know it is there.

Our formulae are wrong, not because we lack the intellect of which science is a part and not because we lack a desire to do what is right. They are wrong, because they are incomplete and we have denied the validity of our spirit as part of our formulae.

The time to find answers is growing short. The ways of the past are too slow for today. Our need is to make sure that people with the vision to know the truth when they see it are talking to the people who know how to implement that truth when it is shown to them.

COMMENTARY

John W. Engroff, Jr.
Adjunct Assistant Professor of History
University of Vermont

We have been concerned with defining or expanding and contracting the definition of pure science as opposed to applied science. In discussing Mr. Conover's paper, however, we can combine pure science and applied science and simply call them science and/or technology. The various distinctions that were made previously are not applicable, because the historical gulf between pure science and technology began to narrow during the middle of the nineteenth century when pure science began to have more technological applications and in fact helped to explain technological processes. In the last fifty years, the time span between pure research and its application has dramatically narrowed. For example, fifty years elapsed in the last century between Michael Faraday's demonstration that an electric current could be generated by moving a magnet near a piece of wire and Edison's construction of the first central power station. Only seven years elapsed between the recognition that the atomic bomb was theoretically possible and its detonation over Hiroshima. The transistor went from invention to sale in a mere three years, and more recently research on the laser beam was barely completed when engineers began using it to design new weapons for the government and new long-distance transmission systems for the telephone company. Theoretical and experimental physicists provided the knowledge out of which hydrogen bombs were made; mathematicians, geophysicists, metallurgists, astrophysicists, and others made the discoveries necessary to construct ballistic missiles; physicists working in optics and infra-red spectroscopy enabled government and corporate engineers to build detection and surveillance devices used in Viet Nam. Consequently, the distinction between pure science and applied science is hardly crucial when examining social responsibility.

Mr. Conover presents a wide-ranged survey of different views of technology which have developed over the last decade. These views which are very sophisticated demand more than a few paragraphs to make them clear. In a sense, even to present them distorts them. The description of the "limits of growth" theory, which has been popular recently, requires additional explication. Mr. Conover suggests its popular proponents have seen only two alternatives: they either assume that "limits of growth" refers to other people without substantially reducing their own standards of living, or they assume that no growth and equal distribution will only reduce the number of their luxuries. The "limits of growth" theory is more complex in its ramifications, as Jorgen Randers indicated recently when he said that perhaps the United States and Soviet Russia ought to stop their growth to allow the developing nations to continue to grow until they were on a par with the "advanced countries."

Mr. Conover fails to deal with two underlying issues involved with science and social technology. One is the use and abuse of power, and the other is the fact that neither science or technology is politically neutral or value-free. C.P. Snow wrote a short book entitled, *Science and Government*, which has not gained the popularity of his *The Two Cultures*. In *Science and Government* he relates the story of a decision

in the British high command made during World War II as to
whether or not to bomb civilian populations in Germany.
Winston Churchill had two very prominent scientific advisors.
Henry Tissard maintained that bombing would not do nearly
the damage as was estimated by F.A. Lindeman who advo-
cated the bombing. Both were distinguished men, but
Churchill accepted Lindeman's advice. The British bombed
Germany quite extensively but without the effect that was
intended. Instead of crippling Germany from the air without
the loss of many British lives, thousands and thousands of
allied air force lives were lost in addition to killing about a
half a million German civilians. Lord Snow told the story to
illustrate the difficulty in getting good scientific advice. If
Churchill had perhaps known a bit more about science and
had a bit more insight into whose advice to accept, Tissard's
might have been taken.

However, it seems to me that Lord Snow missed the main
point of his anecdote. While there may have been good
reasons for bombing Germany in order to shorten the war,
there were greater matters of prestige and competition at
stake on the allied side between the Air Force and the Army
generals. Churchill essentially followed the advice he wanted
to hear. The crucial issue in Churchill's decision was power
and not some gulf between "the two cultures" of humanism
and science. After all, the scientist in government is only a
special case of the expert in government. He has power only
in so far as government leaders listen to him, and he is not
about to weaken his power by telling leaders things that they
do not want to hear.

That science is not politically value-free raises some other
serious questions. Scientific work is done largely by either
corporations and universities which are funded by the government or are tax-exempt institutes funded by corporations.
Although the term, "power elite," is perhaps a bit nebulous,
leaders in the higher echelons of education, government, and
big business do meet freely with one another. Science is done
for particular reasons, many of which have to do with

corporate or governmental interests. "Big science" has social and economic impact. For example, the battle to develop the SST was fought not because of the aircraft's technological merits or because it was good and feasible, but perhaps because of its economic impact in the Western part of the United States which depends heavily on the aircraft industry.

Two important questions have been left unanswered. First, given the fact that science is so intimately tied to power, how does a society safeguard its use? Should there be new regulatory agencies within the government or perhaps a strengthened advocacy system in Congress to balance the powerful scientific advice the President has at his command? Are technology and science somehow intimately tied to the economics of this country in a way that makes them inseparable, and can this relationship of technology and science be changed without in some way altering the basic economic structure? Specifically, capital growth is the *sine quo non* of corporate capitalism, and technology is the motor force of this growth. How can one reform a runaway technology without overhauling the economic system which depends upon it?

Second is the question of a philosophy or the development of a philosophy which will somehow enable us to cope with the perils that technology poses. Can a new philosophy be developed that does not threaten many current values of the corporate establishment? The counterculture is concerned with the earth as a biological entity and as a human entity that can no longer afford to be divided and can no longer support a philosophy which depends upon individualism, competition, and corporate growth. If this concern is in fact part of the emerging new philosophy, how can it be reconciled with the competitive instincts of corporate capitalism, science, and technology?

COMMENTARY

Willard M. Miller
Assistant Professor of Philosophy
University of Vermont

A main theme in Mr. Conover's paper seems to be that science must be viewed as a morally neutral means which may serve either moral or immoral ends. His view of science, with its rigid value distinctions, has had much popularity in our rather positivistic century, because it allows the scientist to justify his endeavors on the grounds that he is not responsible for the way in which others make use of his results. The example of the hammer is instructive; because one cannot know the future uses of one's research, it is therefore not irresponsible to do research and to make the results of it known. However, when one can know the likely future uses of one's results, Mr. Conover argues that the means-and-ends relationship between science and social responsibility is a workable strategy. His statement is somewhat puzzling, since it is not made clear how this strategy is to be understood or employed.

The only socially responsible context for scientific activity occurs when it serves to end suffering and better the quality of life. Without considering those scientific efforts that neither contribute to nor detract from these ends, I want to pursue the question of the extent to which our present scientific efforts are socially irresponsible. The question necessitates a brief analysis of some economic and political institutions that exercise control over virtually all scientific activity in this country.

The United States consumes fifty percent of the world's annually produced resources although it has no more than seven percent of the world's population. Our consumption rate occurs against a background of massive deprivation in under-developed nations where some two billion persons are malnourished and thousands die each day of starvation. Our national consumption of resources is not so disproportionate merely because of our high standard of living; it largely results from massively wasteful forms of production, employed to maintain a rate of growth necessary to our present economic system. To avoid the economic catastrophe that would result from the failure of consumers to absorb the productive output of our continually growing system, style changes, planned obsolescence, convenience products, and credit-buying have been employed on a massive scale. These forms of forced consumerism require staggering quantities of resources far beyond what would be needed by an economy in which production decisions were not based on the profit-seeking impulses of a few people.

To continue our access to resources in vast quantities and to absorb the productive excesses of our system, the United States has the largest military establishment in the world. We maintain our military presence in sixty-four nations in the world, and nearly half of these underdeveloped nations are third world countries governed by totalitarian regimes supported by the United States. Our support is exchanged for the continuance of trade prerogatives which are porfitable for American firms with interests in these countries. Any attempts by the indigenous peoples to liberate themselves from United States domination are frequently met by brutal counter-insurgency programs financed and orchestrated by the CIA. When these measures prove inadequate, as in Vietnam, the United States has shown itself willing to spare no expense in attempting to destroy such movements regardless of the cost in human lives. One searches in vain to find in the Pentagon Papers a single reference to the cost in Vietnamese lives in the plans for the conduct of the war.

The corporate interest also controls our domestic political institutions. As the chief campaign contributors to the two major parties and supporters of intense lobbying activities, corporate management is able to assure the continued complicity of the government in domestic and foreign policy. Domestically, corporations are able to secure continuing access to our resources and the privilege of further employment of ecologically disastrous production practices. For example, within days after Walter Hickel was removed as Secretary of the Interior, off-shore oil leases, which he had refused to authorize, were granted to various oil companies in wholesale lots. The corporate impact on the environment menaces our continued survival, yet our government is unwilling to take any effective measures against it. Our political leaders are loath to bite the corporate hands that feed them.

Whether the American scientific establishment is irresponsible depends upon its relationship to the corporate-governmental structures in such main areas of scientific research as defense, weapons systems for nuclear, biological, chemical, electronic and psychological warfare, and the exploration of space. Of secondary importance are industrial research for the production of new and often unnecessary products with planned obsolescence and useless style change, psycho-sociological research for the sake of manipulating consumer-citizens in stores and at the voting booth, and anthropological studies of other cultures for the sake of dominating them. An example of the latter are the studies of the Mayo tribesmen in Laos which were later used by the CIA to convince the tribesmen to join with the United States in war. These areas of scientific research account for a large portion of the scientific activity of the United States, and it is clear that in these areas, science is used in a socially irresponsible way. The uses of much, if not all, of this research are pernicious ones, contributing to an increase in human suffering and alienation.

It is not enough to state that the scientific community is largely irresponsible; we must ask how the irresponsibility and the conditions it produces may be changed. Notwithstanding Mr. Conover's suggestion that we need to achieve a non-growth condition, we do need to stop growth before we collide with the very limits of our resources. Our present wasteful modes of production are suicidal, but how is a non-growth economy to be achieved? We cannot cling to corporate capitalism in which personal power and profits provide the criteria by which production decisions are made. Further, should economic growth be stopped, it will be necessary to redistribute the wealth of our society, since without growth to support the myth of upward social mobility, we would soon find ourselves in the middle of a class war. Mr. Conover's descriptions of the horrifying consequences following non-growth are imaginary. If we stabilize both economic and population growth, there is no reason to suppose his predictions are well-founded. If we, however, suppose they were well-founded, is he suggesting that we maintain all of these amenities at the cost of our own extinction?

I am not arguing that technology should be eliminated, for to humanize technology by making it meet human needs may be the only way that we as a species will survive. To use technology, we must understand the forces that presently control our technology as well as have an alternative plan for the reorganization of that control. We need to find ways to remove power from those who presently exercise it and make them realize that it is in their own self-interest to give that power to the whole community. We need to complete the revolution we began nearly two hundred years ago; we need to make this nation a genuine democracy.

Our present absorption in competitive, self-indulgent individualism has resulted in our existence being continually menaced by the real possibility of thermo-nuclear and environmental disaster. Only by developing cooperative modes of human interaction are we likely to survive. We must humanize corporations and let them be controlled by people who work in the communities they affect.

Mr. Conover suggested that the conflicts between corporate attitudes and idealistic humanism might be resolved by discussion. At best this suggestion is naive; at worst, it is an attempt to obscure a fundamental conflict between the demand to fulfill the organization's responsibilities and the need to fulfill one's moral responsibilities. When the idealistic student is asked to follow orders for the sake of the corporation, he is confronted with a question of authority and power which is not likely to be answered in a *tête-á-tête* over tea.

To a large extent the scientific community is a handmaiden to the most socially irresponsible forces. Either scientists must dramatically reassess their roles from a moral point of view, or they must consider themselves as a part of what menaces us all. Scientists like Enrico Ferme, Leo Zelard, Robert Oppenheimer, Noam Chomsky, and Daniel Ellsberg have taken morally responsible stands against business-as-usual in science. Their stands are made at the costly expense of their careers, yet in an important sense theirs are not courageous stands, they are only moral ones. That simple morality appears courageous is a sign of how desperate modern living has become.

There are grounds for cautious optimism, for in numerous scientific and professional associations, voices are to be heard in ever greater numbers, calling for scientists to take a stand by no longer abrogating their roles as moral agents. If and when scientists join others moving toward fundamental social change, they will be moving toward a commitment to full social responsibility. In the end science must serve the needs of all passengers on "spaceship earth." If it does not, I fear we are on a very short voyage.

COMMENTARY

Stanley M. Grubin
General Manager, Service Division
Western Electric Company

I would like to make two value judgments. First, man is becoming less and less likely to sit back and allow either the literary scholars or the scientists to establish the goals of society. Secondly, change is being brought about so rapidly that unless solutions to problems can be developed and implemented almost instantaneously, the definition of a problem will erode during the problem-solving stage.

A need has been expressed to bring the literary scholars, the philosophers, and representatives of other disciplines related to the humanities closer together with scientists in order to determine society's direction and the strategy for best achieving that direction. We seem to be saying that Lord Snow's two cultures, which really represent a very small portion of society, should meet, "slice up the pie," and dictate goals and methodologies. I do not think the rest of society, as it becomes increasingly sophisticated, will accept the dictates of a few. A recent comment on British society implied that too many people had achieved a level of education beyond their social and economic class. Is it not likely that the increased restlessness on the part of the populace can be partially explained by their unwillingness to allow others to set their goals?

Recognizing this possibility, people in industry, with the help of social scientists and philosophers, are experimenting with approaches to motivating employees. One promising, recent approach is called, "job enrichment," which allows employees certain freedoms in accomplishing a job task. The potential importance of the "job enrichment" concept was dramatized at a recent seminar, "Central Influences on American Life," sponsored by the National Commission on Marijuana and Drug Abuse. Among the conferees were Jonas Salk and Jay W. Forrester. One theme kept recurring: for society to be stable, peaceful, and happy, people must feel satisfied and fulfilled. They must feel as if they contribute substantially to society as a whole, and further, they must be able to see how their part affects the whole. One way to achieve meaningfulness and fulfillment in a job may be to have employees participate in creating the structures of their jobs.

Excluding employees from participation in developing job structure—heretofore a traditional and acceptable procedure—can now result in a situation similar to the shutdown of the Vega automobile production line at the General Motors plant in Lordstown, Ohio. The Vega administrators in Ohio discovered that on their production line, like other assembly line operations, many individual workers were increasingly unable to relate their work to the whole, the assembled automobile. Their jobs, therefore, tended to lose their meaning; the productivity of workers suffered. Workers today are questioning more and more the output of their labor, and the impact of the output of their labor on society. The key words are "output," which is determined partly by technology and partly by motivation, and "impact," which is measured by the effect of the output on this and future generations.

If we accept Lord Snow's thesis that there are two cultures

and believe that philosophy asks why and gives directions while technology determines how we move in one direction or another, then not only are the two cultures going to have to talk with one another but they are also going to have to talk with the rest of the world. Since the individual in society is increasing his demand for a role in determining society's direction, we should discern what direction he is *seeking* rather than dictate that direction to him. In determining social values, the philosopher may have to talk with assembly line operators to determine the "why" and the "what." The scientist may have to rethink automation to determine the "how."

We are going to have to reconsider our traditional approach to evolving change, which typically consists of defining a problem, solving it, and implementing the solution. Our approach will have to be streamlined greatly, because the solving and implementing require so much time that we may solve the problem as it existed when we confronted it and not as it exists today. Even our legislative process might have to be modified so that ecological matters could best be resolved by having instantaneous referenda.

RESPONSE

D. K. Conover

I must confess to some bewilderment regarding the commentaries of Professors Engroff and Miller. On one hand, they reject the proposition that science or scientists may be "value free", i.e., employed for good or evil beyond the power of the scientist to decide. On the other hand, casting science in the role of handmaiden to powers clearly outside the scientific community, they insist some new form of controls are necessary to direct science along paths of higher morality. Like Mohr's circle, they have started on one side and ended up on the opposite.

This is not logic but rhetoric. Moreover, the rhetoric seems aimed not at science or social responsibility but toward a political ideology. Professor Miller apparently feels that talking about conflicts of power and authority are pointless. He takes my reference to the very real problems of the recent graduate entering industry and turns it into a . . . "*tête-á-tête* over tea". Even the people he seems to admire, he disparages with the odd notion that because their courage stems from morality, it is somehow less important. This is indeed a difficult ideology to understand, and one feels uneasy about the harsh judgments and the prospects of hard treatment for transgressors. In the absence of any empathy with those of us who fail to perceive conspiracy around every corner, what comes through of Professor Miller's politics sounds like a threat.

It is also rather sad because in choosing to make this a contest of ideologies, they have said nothing about understanding or compassion, or love. Both Professor Engroff and Miller seem to agree that any person or institution possessing power is necessarily corrupt and, apparently, beyond redemption. This, of course, is their right; however, as I said before, it is rather sad.

If opinions are important, I might as well add my own. It is based on a limited view, my own experience. I have worked as an engineer and as a manager in a large corporation. I know scientists, and I know corporate executives. Occasionally, I have seen them make the kind of socially significant decisions we have been talking about today. In a small way, perhaps I have done so myself.

From such experience, two things have impressed me over and over again. The first is how hard it is to know what is right when you have to act, not just talk. There is always something you don't know, some uncertainty that makes each possible choice a question mark. The second thing is that there is usually someone who looks at the same situation and reaches an entirely different conclusion. Not just a different assessment of facts but different conclusions about the moral principles involved.

What remains to be concluded is one's opinion about the wisdom and goodwill motivating such diverse positions. In my presentation, I attempted to characterize certain of the perspectives that lead different people to opposing views. If the issue were simply that some people are good and some are bad, this entire discussion is meaningless. It is only by recognizing that logic and sincerity based on a particular, limited perspective leads to legitimate argument about ends and means that we have a situation that might be solved rationally.

Obviously, evil exists. Every human endeavor is punctuated by the existence of charlatans or criminals who abuse the rest of us. But in my opinion, such people are a minority, and whatever power they possess is not enough to match the many decent people, in every walk of life, struggling to match humane instincts with appropriate actions. If this were not so, I don't think humanity would have survived this long.

On the contrary, man's capacity to feel compassion, to love, to learn the ways of justice and equality, is the motivation of real progress. Sometimes we learn slowly, but the course of history, as John Kennedy said, " . . . is on the side of liberty." The institutions we now have are far from perfect. But, they are dynamic, and I believe they are changing for the better.

To the extent that Professors Engroff and Miller are impatient in their desire for improvement, so am I and so are most of us. But to deny that progress has been made, or to castigate an entire social system because it is imperfect, strikes me as unwarranted and extreme. Moreover, their failure to provide any clue as to the structure or philosophy of an alternative verges on irresponsibility.

So much for differences of opinion. If our purpose is to inform so that the people can make up their own minds, (which they surely will do anyway), permit me to return to the specific issues. Those I want to mention briefly are the limits to growth debate and the question of corporate power.

Professor Engroff has said that presenting the different views of technology which have developed is so complicated as~to distort them. Professor Miller has said our technology has been co-opted by the military and profiteers who manipulate consumers into a life of conspicuous consumption and waste. Although I believe they have overstated the case, there is certainly some truth to what they say. However, these are value statements about how we use our resources, and the limits to growth study is more about the finiteness of resources available regardless of how they are used.

The real issue in limits to growth is what to do about finite supplies in a world of expanding demand. Professor Miller implies that if Americans would only stop wasting so much, if we would forego the right to make personal choices about consumption, and if we would redistribute the wealth of our society, we could stabilize population and economic growth. He warns that failure to do all of these things is suicidal.

As I attempted to point out earlier, the gap in wealth is so great and the affluent of the world such a small minority, that the matter goes far beyond what we waste. To do what Professor Miller seems to be advocating would have little effect other than to destroy the social and technological base from which a real solution might develop.

I agree that we need to stop wasting both people and resources. I hope for the kind of world where military expenditures can be eliminated. But, to achieve these and to provide real help to the billions of people living at the edge of survival will not happen as the result of our own self-destruction. Without affluence we would have neither the education nor the time; indeed we would not be able to conceive a study such as the limits to growth. And, we would certainly lack the means to discover a solution.

Part of any solution that works will involve organization. That is the way man gets jobs done, the way the world is. And part of what makes an organization work is power; that is also the way in which the world works.

The practical questions about power are what kind and how to control it. Professor Engroff complains that Winston Churchill was not a better scientist. Is the implication that we should seek an all-knowing leader, a philosopher king? Such a person would be wonderful to have. Organization would be less of a problem, and we would benefit from perfectly integrated knowledge and wisdom. Until such a king made his first mistake. Then we would call him a tyrant.

Professor Miller wants power reallocated to something he call, "the whole community." This is a worthy ideal, the

practicality of which depends on our ability to retain the
necessary elements of organization. Without a plan for
organization, the ideal degenerates to anarchy and paves the
way for another form of tyranny.

In contrast, Mr. Grubin recognizes man's desire for more
individual control and, at the same time, our lack of easy
answers to solve the problems involved. He calls on people of
diverse talents to work together to find solutions. He recognizes that time is short. We may not be able to wait for an
integrated philosophy, a new socializing principle.

We need to get moving. Describing the "job enrichment"
concept, Mr. Grubin is talking about a step which is reallocat-
ing power to more of the "community." Instead of revolution, he is describing evolution. It assumes some faith in
people, those with power to share, and those stepping up to
new responsibilities. From my view of history, he is talking
about a way that works.

I would like to end this response by harking back to the
end of my paper in which I made a plea for our spirit,
for something to lift us a little bit higher than we have been
before. All of us have been talking about improvement and
progress even if we disagree about some of the particulars.
None of us wants to retreat. Some of us will cling to our
Toledos, but we are all happy to leave the Inquisition to the
past.

BIBLIOGRAPHY

William R. Burch, Jr., *Daydreams and Nightmares, A Sociological Essay on the American Environment* (New York: Harper and Row, 1971)

Jacques Ellul, trans. by John Wilkinson, *The Technological Society* (New York: Alfred A. Knopf, 1964)

Victor Ferkiss, *Technological Man: The Myth and the Reality* (New York: George Braziller, 1969)

Donald P. Lauda and Robert D. Ryan, editors, *Advancing Technology: Its Impact on Society* (Wm. C. Brown Company, 1971)

Herbert Marcuse, *One-Dimensional Man* (Beacon Press, 1964)

Noel de Nevers, editor, *Technology and Society* (Addison-Wesley Publishing Company, 1972)

Jerome R. Ravetz, *Scientific Knowledge and Its Social Problems* (Oxford: Clarendon Press, 1971)

Albert H. Teich, editor, *Technology and Man's Future* (New York: St. Martin's Press, 1972)

Science and Technology: Tools for Progress Report of the President's Task Force on Science Policy (Washington, D.C.: US Government Printing Office, April, 1970)

AFTERWORD

A FINAL RESPONSE

Henry John Steffens

This afterword is intended to explore some of the problems, issues, and contentions raised by the preceeding papers and deserving of further discussion. Its purpose is also to extend some of these concepts to new areas. These new areas will include discussion of 1. the self-limitations and hidden assumptions in our modern "myth" of objectivity; 2. the possible positive consequences for the biological sciences and for our own conception of human life which might follow the relinquishment of the power of this "myth"; 3. the possible expansion of orientation and world-view beyond mechanism and nineteenth century determinism; and 4. the failure of social responsibility arising from the confusion between the activities of knowing and understanding.

A narrowness and closed-endedness seems to be characteristic of any well-established system of thought. This is as true for the sciences and technology as it is for established modes of political, economic, or social thought. An inertial aspect to thought processes "acts" to resist changes in direction and speed. Perhaps the broadest example of inertial thinking is the confidence Western society has placed in objectivity. Ever since the Scientific Revolution in the seventeenth century and the successful Cartesian establishment of the mind-body duality, Western society has slowly succumbed to the belief that the rational, analytical approach, confirmed by repeatable experiment and controlled observation, is the only viable access to knowledge of the physical world. Implicit in the assertion of the mind-body, spirit-matter dualism was the implication that the world of matter was the only "sure" world, and certainly the only one susceptible to analysis and mathematical expression. The world of spirit was considered suspect, in large measure because of its inaccessibility to the "objective" approach used in dealing with matter. Confronted with this either/or choice between spirit and matter, the West seized upon matter and relegated spirit to the "fuzzy-headed" areas of philosophy and religion. As the nineteenth century progressed, religion was shunted into a side channel of the mainstream of Western thought, and philosophy, in general, attempted to draw closer and closer to the model of science. In brief, for Western society in what has been called the "Age of Materialism," scientific method and objective statement were "in"; intuition and subjective response were "out."

This societal preference was reinforced, as the papers have suggested, by a wide variety of factors. Three important ancillary assumptions, covert in this preference, have recently emerged into the open for public scrutiny by the events of the 1970's. The assumptions are that: 1. all aspects of our world, and all observable elements therein, are susceptible to complete and objective analysis; 2. all problems, once recognized and described, are capable of solution; and 3. modern technological man, armed with scientific method and technological techniques, can and will eventually solve all problems with his previously successful methods and techniques. Seldom in this confident and optimistic orientation does there appear to be room for serious rememberance of the other half of the dualism—spirit, and seldom does there appear to be much serious consideration of possible incompetence stemming from the very methods and techniques so confidently employed. Yet, current problems—among them the Indo-China War, the problems which accompany current modes of political practice in the United States and abroad, and the problems created by the world-wide use of natural resources by modern technological society—all suggest that something indeed needs to be corrected in the assumptions and preferences of Western society.

We seem recently to have come to a turning point in public awareness of the need for modification and improvement. It is no longer feasible to abdicate public responsibility for directions in society because of a claimed lack of ability to employ a group of very specialized methods and techniques. The possession of scientific method and technological technique by experts can not, by itself insure meaningful solutions to widely recognized problems. The results of the schemes of the "best and the brightest" now seem incompatible with any publicly recognized sense of meaningful activity and appropriate results. The public's confidence in the expert with his expertise, his "state-of-the-art" information, and his tacit belief in the three assumptions just mentioned, has begun to erode.

This erosion has begun because the experts' claims of solving current problems have not produced the expected results. This failure of expectation did not occur because the knowledge, techniques, and methods which the experts provided were faulty in themselves, but rather because their expertise failed to convey meaning. The experts' facts do not speak for themselves, but must be given meaning and context. This meaning and context, in turn, provides the possibility of realizing the limitations of the data. Many now realize, moreover, that we cannot sensibly ask the military man, the politician, or the oil executive to provide the "objective" meaning of raw data which they themselves have collected or were responsible for collecting. This realization has much less to do with any given individual's honesty or willingness to provide meaning than it has to do with his state of captivity by his accustomed system of usage and practical concern. An expert's meaning can become captured by the organizational structure behind the collection of the data, and his modes of expression can be shaped by the language used. Meaning becomes tied into process and practice and is often shaped prior to collection by the attitudes underlying the process of collection itself. On a simple level, for example, when confronted with the "fact" of a milk bottle with a half of its internal volume containing milk, an "objective" observer can claim that it is half-full or half-empty, depending entirely upon his predisposition and his reasons for taking note of the bottle in the first place.

It is not possible to introduce fundamentally different innovations by playing the old game according to the old rules. The "energy crisis," for example, will not be solved by re-packaging old notions into different conceptual containers and by reasserting the same terminology with the same hidden assumptions, preconceptions, and predispositions. New solutions require new rules, new sets of assumptions and values, and usually, new controllers of the game—if not actually new players themselves. It is, in some sense, unfortunate (though certainly arguments can be made for conservatism and the status quo) that people involved in active modes of life are seldom aware of their preconceptions and predispositions. The myth and goal of objectivity is too deeply ingrained in Western society to make it anything less than extremely difficult for us to successfully separate our irrational, subjective commitments from what we insist upon as thoroughly rational, objective activities. We have seen, too, that it is now possible to abandon quickly a mode of action once we have embarked upon it with what we were convinced were the best of reasons. The Indo-China conflict, our attitudes towards the use of natural resources and energy, and the insistence upon rather simply expressed solutions to extremely complex and incompletely understood social problems have loomed up to attest to our inflexibility. It is extremely difficult, for many reasons, to turn aside from a path of action once we have committed ourselves to it. Indeed, the usual technique in our society has been to offer every possible justification to maintain the action along a previously chosen path. If, under extreme duress, the status quo cannot be maintained, slight modifications of current practice are acceptable, as long as these modifications bear a rational tie to the accepted path of action.

The adherence to the myth of objectivity in the sciences has produced a host of somewhat surprising assumptions which have affected both the sciences and our confident attitudes towards society. Perhaps the central theme generated by this myth is a commitment to the belief in the possibility of separating the observer from that which is observed. This commitment produces the confidence that any subject can be scrutinized and analyzed objectively. It also leads to the tacit assumption that these objects will lend themselves to specific and unchanging statements and solutions. It further suggests that the laws of activity, compiled from isolated observation, will be established as true and will remain true and unchanging for all time. This orientation places a premium upon methodological ability and process rather than upon free creativity and expression. This stress upon technique and virtuosity is only a symptom of a more important imbalance in our societal orientation.

We have chosen to revere objectivity modeled on the old, outdated image of science at precisely the time when the old model of objectivity was being shown to be inaccurate. The achievements of the modern physical sciences—both relativity theory and quantum theory—have demonstrated that there can be no absolute frame of reference, no position of observation of, or no pronouncement about the world which can be separate and separable from the world observed. The nineteenth century confidence in the careful, removed observer making "true" statements about a world which he observes objectively and to which he brings some absolute frame of reference has been challenged by modern physics.

This challenge has made many aspects of our old world-view stand out in stark, and to some, rather embarrassing relief.

There has been a variety of responses to the increasing modern awareness that the model of objective statement, carefully developed and nurtured during the nineteenth and twentieth centuries, could not be maintained. The responses range from existential despair to a new feeling of freedom to pursue other areas of investigation. Artists, composers, poets, authors, and, in general, members of what Lord Snow termed the "literary intellectual community" have generally responded with essentially negative expressions and rather sustained pessimism. Some, having chosen to renounce all hope for meaningful statements, have isolated themselves from society. Others have reconciled themselves to isolated, personal expressions of experience and have attempted to provide raw experience without communicating meaning.

To give up the goal of meaningful expression is to fall prey to the same mistake which led modern men into believing the myth of objectivity in the first place. Because the Western world has chosen to revere the model of objective science and its method, process, and technique, it does not follow that 1. the choice was correct because science and technology have been successful, or that 2. the spiritual values and communicable meaning of human life which this reverence has rejected were not there to begin with, and indeed, are not still there. The adoption of a system of orientation or thought which degrades and eliminates the spiritual, the subjective, the intuitive, and the non-verbal does not mean that these alternatives are indeed not important. Denying the aspects of human life which do not fit into the formalism of our currently accepted modes of expression does not mean that these aspects of life do not exist. It means only that we have convinced ourselves that there really is only one approach to the world. A rejection of the dominance of the empirical approach—the objective model—can open the possibility for the re-emergence of an appreciation of subjective, intuitive aspects of the world, and for the possible restatement of mutually held values and communicable beliefs. These possibilities are as true for new areas of the sciences as they are for new modes of expression in the arts and new attitudes towards politics and towards society.

In the biological sciences, a serious consideration of better defining the limits of applicability of the mechanistic view may open up heretofore rejected areas of life and activity for scientific consideration. This scientific consideration in the biological sciences will have to accommodate modified notions of what appropriately constitutes scientific investigations. At least three important areas require modification: 1. the inclusion of concepts from quantum mechanics to modify the mechanistic position in biology; 2. the possibility of inclusion of subconscious activity into the realm of scientific consideration; and 3. the accommodation of subjects which are outside current scientific acceptability—subjects such as extrasensory perception, psychokinesis, inheritance of acquired characteristics, and the concept of mind.

The triumph of the mechanistic approach to the biological world is of more recent origin than most biologists like to recall. Vitalism, while consigned to obscurity, was by no means vanquished. Neglect and the hope that it would somehow disappear did not constitute sufficiently compelling arguments to abandon it. There were, to be sure, practical reasons for ignoring vitalism in biology, as there had been reasons for ignoring metaphysics in the general progress of science. Stress on the physiochemical aspects of life allowed biology to become more rapidly "scientific," more easily modeled after the standard of scientific objectivity proposed by the physical sciences in the nineteenth century. The mechanistic approach in biology stressed those things which could most easily be subsumed under quantitative description. This approach made striking and unquestionable progress in revealing a tremendous wealth of information about biological processes, physiological functions, and behavioral characteristics. Mechanisms were conceived and supported throughout the whole range of the biological sciences; the

DNA double helix is only one popular example of the tremendous range of subtle and complex biological processes accounted for by equally subtle and complex mechanisms. Vitalism, on the other hand, proved a hindrance to the biological sciences. It marked off areas which seemingly could not be known but which mechanism ultimately explored. As a result, the mechanistic approach dominated while vitalism was degraded and forgotten.

With the demonstrable success of the mechanistic approach to biology, it became convenient to overlook those areas of deep, continuing interest which were not covered by the rigorous demands for objectivity of this approach. During the 1940's, 1950's, and 1960's these areas could be conveniently avoided, for the most part, because so much else needed to be done. Ironically, however, as mechanism reached its most mature stage of development, the need for a form of vitalism began to re-emerge. By the 1970's it was becoming increasingly clear that although biological mechanisms could offer an explanation to virtually all biological functions, a few remaining areas outside the context of widespread investigation seemed resistant to the previously successful methods. A renewed interest in vitalism, with its insistence that life processes are unique in character and different from chemical processes, offered the possibility of a broadened approach which would encompass these remaining areas of investigation.

Molecular biology provides an example. For many years, mechanistic interpretations provided the most exciting areas of study for molecular biologists and produced several noteworthy advances, including the "cracking" of the genetic code and the assertion of the "central dogma" with its insistence upon the stability of this code. Many, however, now recognize several areas in molecular biology which, though assumed to be susceptible to the mechanistic approach, seem more clearly aligned to the central issues of vitalism. These areas include: the process of cell differentiation, the origin of life, and the functioning of the central nervous system, especially the human brain. Mechanism seems unable to provide answers to questions concerning the purpose of life and the nature of willful activity and consciousness, and scientists have only recently admitted the possibilities of extra-sensory perception and areas of investigation broader than, and perhaps inclusive of, physiochemical mechanisms.

Many indications of the limitations of the mechanistic orientation exist. The earliest is derived not from biological sciences but from physics. Two of the founders of quantum physics, Erwin Schrödinger and Niels Bohr, postulated the limitations of mechanistic biology. Schrödinger, in his essay *What is Life?* suggested that the limits to knowledge posed by the Uncertainty Principle would prevent biologists from ever probing the mysteries of the cell. Modern molecular biologists have demonstrated, however, that they can perform experiments in vitro which can unlock these activities on the molecular level. They thus have rather successfully avoided a direct confrontation with the limits to biological knowledge posed by the Uncertainty Principle. Improvements in techniques of modern molecular biology seem to have successfully met the first limitation by the interesting technique of removing the study of life from life. Niels Bohr, however, pointed to the limitations of the physiochemical processes themselves. He used his concept of Complementarity as his source of inspiration. He asserted that some aspects of life—for example, consciousness—could not be fully analyzed by the techniques of mechanistic investigation. That approach, he argued, could not encompass the whole process at once but had to rely upon the analysis of separate parts. Once any one aspect of consciousness was isolated for analysis, it became quite usual and subject to physiochemical accounting. However, when the analysis of each part was completed, the investigator was left with a group of assembled facts which did not produce a whole understanding of consciousness. Bohr was quite pessimistic over the prospect of ever understanding consciousness. The physicist, Victor Weisskopt, has recently expressed Bohr's despair as follows:

The awareness of personal freedom in making decisions seems a straightforward factual experience. But when we analyze the process, and follow each step in its causal connection the experience of free decision tends to disappear . . . Bohr, an enthusiastic skier, sometimes used the following simile, which can be understood perhaps only by fellow skiers. When you try to analyze a Christiania turn in all its detailed movements, it will evanesce and become an ordinary stem turn, just as the quantum state turns into classical motion when analyzed by sharp observation.

The molecular biologist, Gunther Stent, has suggested a similarly pessimistic viewpoint:

This attitude would mean nothing less than that searching for a "molecular" explanation of consciousness is a waste of time, since the physiological processes responsible for this wholly private experience will be seen to degenerate into seemingly quite ordinary, workaday reactions, no more and no less fascinating than those that occur in, say, the liver, long before the molecular level has been reached. Thus, as far as consciousness is concerned, it is possible that the quest for its physical nature is bringing us to the limits of human understanding, in that the brain may not be capable, in the last analysis, of providing an explanation of itself.(1)

This pessimism is a mistake, the same mistake as evidenced in our previous discussion of the rejection of spirit. That the current methodology does not yield the anticipated understanding does not mean that that understanding is not possible.

This attitude of pessimism is both correct and fundamentally wrong. It is correct in that the lessons of modern quantum physics should by now be clear. A limit to the extent to which we can obtain exact information about certain events exists in both the physical and biological sciences. The observer must be taken into account, especially in the world of the very small. There is no possibility of completely detached observation, either of electrons or of certain biological functions. It does seem apparent that the current analytical path of access to the sub-atomic realm, or to the realm of consciousness, has been blocked by fundamental limitations which cannot be circumvented by more sophisticated instrumentation and techniques. The analytical path however, is not the only path. The principle of Complementarity itself suggested what is now becoming increasingly clear in practice: one view, one direction of investigation, or one approach to knowledge should no longer be expected to reveal the complete truth. This is as true for modern biology as it is for modern historical scholarship, for the physical sciences, for the social sciences and for the whole range of intellectual endeavor. Because the world-picture generated by the myth of objectivity has been so successful in so many informational and material ways, it does not follow that the myth or the world-view it generated is true for all time. The alliance of "objective science" and Cartesian mind-body dualism has almost convinced modern society that the myth of objectivity is not only correct, but constitutes the only approach to true knowledge of the physical world and of ourselves.

It would be foolish to deny the tremendous success of modern investigations employing scientific methodology. But it also seems foolish to ignore the information available from the history of science, which suggests that this so-called objective process is often permeated with hidden assumptions, sub-conscious imagery, and non-verbal insight. Rational, orderly, logical, well-ordered verbal thought does not represent either the whole range of the process of creative thought or the whole capabilities of human abilities. Contrary to the impression created by the exclusive emphasis upon rationality and objectivity, allowance for the importance of the sub-conscious, the non-verbal, and the subjective in science does not cast us into a quagmire of obscure statement. Rather, the serious consideration of such factors will allow for new possibilities for investigation, and will open up broader vistas. If past experience can be used to help interpret possible ranges of action in the future, the older mode of thought will prove tenacious and will resist replacement by those who have espoused it with great persistence. A new, broadened perspective will most likely encompass the old, offering a different world-view while including the old formalism within its newly devised formal statement. Relativistic physics is an example of a broadened position which includes the more limited classical formulation. The new simply does not build on the old, thereby producing a broader perspective. The new introduces a changed world-view, in which the old is seen to be both limited and isolated. The old view is seen to have hidden previously unappreciated assumptions and conceptual difficulties and is no longer viable as a part of the foundation of a new world-view. Certainly the data generated by the old system, such as classical mechanics, were not "wrong," just as the new mode of approach, relativistic physics, is not now "right." However, the results produced by the old system can be perceived from a new perspective, as part of an understanding of the world limited by assumptions which were appropriate but are no longer viable. They cannot be maintained when confronted by the new conceptuality and the newly generated evidence for a new mode of thought.

How this new mode of thought develops, of course, is the problem. Arthur Koestler has succinctly summarized a view of the act of creation as follows:

> . . . ordered, disciplined thought is a skill governed by set rules of the game, some of which are explicitly stated, others implied, and hidden in the

code. The creative act, in so far as it depends on unconscious resources, presupposes a relaxing of the controls and a regression to modes of ideation which are indifferent to the rules of verbal logic, unperturbed by contradiction, untouched by the dogmas and taboos of so called common sense. At the decisive stage of discovery the codes of disciplined reasoning are suspended as they are in the dream, the reverie, the manic flight of thought, when the stream of ideation is free to drift, by its own emotional gravity as it were, in an apparently 'lawless' fashion.(2)

An objective, analytical and verbal approach does not allow for the serious consideration of subconscious activity. It precludes not only subconscious "thinking," but also the possibility of subconscious motivation or the influence of the subconscious upon health and bodily function. Stress upon conscious activity, rationality, and strict causality in observable events, can begin to be viewed as a highly successful, but increasingly limited, orientation for modern society. An interesting example of such a limitation can be derived from modern research into the questions of the nature of the mind, of consciousness, and of the relationship between mind and matter.

Evidence has become undeniable in recent years that a phenomenon called Extra Sensory Perception exists. The evidence indicates that there "does exist a small number of people who obtain knowledge existing either in other people's minds, or in the outer world, by means as yet unknown to science."(3) Controlled experiments have yielded results which could not be expected on the basis of probability occurrences. The larger the number of observations taken, the greater is the certainty that the extra sensory phenomenon under investigation is something other than a random happening. (It is not unusual to obtain results of one million to one against the chance that the findings of the experiments are attributable to luck.)

Experimental results of this order of certainty would have been accepted with high confidence if they had concerned some recognized physiochemical process. However, because the area of investigation challenges the myth of objectivity and the Cartesian separation of mind and matter, experiments with ESP, and the even more unusual investigations of psychokinesis, have been only slowly and reluctantly considered acceptable for scientific investigation. Experiments in parapsychology seem to call into question our concepts of causal relationships in time, the nature of matter, and the nature of space occupied by human beings. That these concepts have already been challenged and modified by the physics of the early twentieth century does not seem to have shaken the willingness of the biological and social sciences to cling to their old notions and preconceptions. It is, in large part, the old predispositions which have made it so difficult to take seriously the search for a replacement to our current biological and social understanding. It should be emphasized again, however, that the data and factual information gathered by the biological and social sciences are not in question. The meaning of the data and information and our current understanding of the world is what is being questioned.

What, after all, can the old view of space, time, and causality make of experiments in psychokinesis. If it seems possible for the mind to know things outside of the usual processes of knowing, why is it not possible that objects may be influenced by means not currently considered feasible? Perhaps mind can influence matter directly. Recent experiments with dice and with radioactive decay of strontium 90 indicate rather convincingly that seemingly random events may be influenced by voluntary human effort. Helmut Schmidt, as director of the Institute for Parapsychology at Duke University, produced a series of experiments on precognition and psychokinesis. The experiments yielded results which mitigated against luck by odds of the order of thousands of millions to one in predicting the outcome of theoretically unpredictable sub-atomic processes.(4) Extra-sensory Perception, precognition, and psychokinesis are now increasingly acceptable areas for serious scientific concern. Their successful investigation will depend upon the investigator's ability to transcend current reasonable expectations and to allow serious consideration of what may now seem far-fetched and bizarre.

The point in this Afterword is not to illicit support for parapsychological research, but to stress the concept, raised by the papers, that our orientations and accepted actions need to be changed. These changes seem increasingly necessary in the light of recent events and return us to the issues of social responsibility raised in the last paper. As it was insufficient to rely upon modification of the currently acceptable to produce what is broader and new, so, too, it is insufficient to continue to assert current notions of social responsibility without a fundamental reassessment. Insisting upon the old with increasingly greater vehemence does not make it new. Nor does the repeated call for increasing awareness and understanding, within a context which de-emphasizes that same awareness and understanding, yield the hoped-for result. The papers sufficiently clarified the issues supporting the assertion that insistence upon "more of the same, just better" does not provide an environment which nurtures creative reorientations and new understandings. It is not appropriate to demand that people engaged in processes and methodologies which have produced results of questionable social value "reform" themselves and become more socially responsible in the future. The problem is, and has been, that their techniques and methodologies do not yield values. Such values must be realized from other sources, and these other sources, outside the productive processes of a technological society, have been under a rather extended eclipse.

Calling on scientists and technologists to be more socially responsible in the future is misguided, for at least two major reasons implied in the papers: 1. science is not a process conducted according to established methods which yield predictable results, and 2. technology is founded upon established methods designed to accomplish recognized objectives which do not emerge from the technological process. Simply stated: scientific creativity does not know where it is going, nor what new understandings it will produce; it cannot therefore be required to produce socially responsible results in areas it does not yet understand. Technological processes cannot be used to produce socially responsible results, because the technologies are employed to achieve goals which have already been envisioned and determined. To demand responsibility from processes which do not produce or include values is unrealistic. It seems equally misguided to demand extraordinary, socially responsible efforts from precisely those persons who have immersed themselves in fields which are unrelated to social concern. They have little training and little time for awareness of social questions.

Social responsibility is the responsibility of every member of society. Scientists must be expected to act responsibly because they are members of our society, not because they are scientists. Technologists must likewise be expected to behave responsibly, not because they are technologists, but because they are members of a community of men. Values and goals must be provided and enforced by society as a whole and not just by scientists or technologists. The current abdication of responsibility on the part of the public to learn about, and understand, science and technology takes on more than just a realization of ignorance; it points to the source of the failure of social responsibility. Recent world events have demonstrated that society cannot disengage itself from technological, political, and social processes and leave them to experts, if society expects to be able to obtain results which conform to values having no natural place in those processes. People and groups of people formed into societies agree upon values: it is their responsibility to see that they adhere to these values and augment them by action. Abdicating this responsibility of persistent monitoring, and educated, informed criticism has been an all too frequent characteristic of twentieth century society. Public awareness has abdicated to expert specialty. The argument is frequently voiced that the modern world is so complicated that no one can understand what is happening in all areas of it. Moreover, it continues, our best hope is to develop expertise in one specialized area, and let experts take care of those other areas (most of the world) outside our self-limited zone of competence. This approach confuses knowledge with understanding, and it employs the mistaken argument which Lord Snow called the "two thousand and two cultures" approach.

The argument that the world is too complex for any one person to understand is usually linked conceptually to reluctance toward seriously entertaining the idea of fundamental change. As Lord Snow put it in his *Second Look*:

> This attempt at excessive unsimplicity, the 'two thousand and two cultures' school of thought crops up whenever anyone makes a proposal which opens up a prospect, however distant, of new action. It involves a skill which all conservative functionaries are masters of, as they ingeniously protect the status quo: it is called 'the technique of the intricate defensive.'(5)

The 'technique of the intricate defensive' seems to lie at the base of abdication of the individual's responsibility to see the directions and goals of the processes of his society. By claiming that there are many more than the two cultures which Lord Snow delineated, the argument loses sight of the definition of culture which Lord Snow used to define his argument. He spoke of culture in the sense of "a group of persons living in the same environment, linked by common habits, assumptions, a common way of life." Culture, obviously, can be defined in many different ways, but if discussion is to take place on the "two cultures" theme, "culture" can only be used to indicate some collective characteristic. Emphasis upon the increasing fragmentation of our society into specialized units would indeed support the insistence upon the proliferation of the number of the cultures. In the extreme, the paper on the two cultures in this book used the concept of an "infinity of cultures" for the sake of argument. Of course, even if we were to assume that there was one culture per human being, the number of cultures could in no way be infinite.

Stress upon fragmentation lends support to the necessity of pooling expertise in the hope that all important areas of specialty are covered by some specialist. Research teams are emphasized to insure a broad reservoir of information and technique. Specialization retains the separation of cultures and, indeed, insures the continuation of separation because each unit of specialty is defined by areas of technique and expertise. Hope for "bridging the gap" between separated fields of endeavor is meager, because, as has been well recognized, no effective mode of communication exists between dissimilar techniques and methods. The road to unity lies in a different direction emphasizing understanding, in addition to knowing.

It has been suggested in the second paper that the problem of the two cultures can be dismissed by claiming the de facto existence of many more than two cultures which are a positive blessing that "is central to the evolution of civilization." This suggestion implies, as Lord Snow indicated, that the current condition should be maintained in general outline while we discover new techniques and methods to improve our world. Mr. Cook suggests a new method of societal improvement—the "politics of creative tension"—and recommends that we apply the "systems concept and method" to meet our current problems. He refers to "think tanks" and suggests that it is through the more ubiquitous application of "systems engineering" that the road to a better civilization lies. The assumption, of course, is that our problems are amenable to solution by the use of proper team research techniques and systems design methods. Mr. Cook concludes with the remark, "we are all building (Sir Francis) Bacon's pyramid of knowledge."

Surely no one can doubt the tremendous past successes of the approach Mr. Cook has outlined. But, conceptually, there can be a very different way of viewing our current situation, which places this approach in a secondary, but still important, position. Perhaps our technological society should try to escape from under the "pyramid of knowledge" so that we can more meaningfully stress understanding rather than virtuosity of technique. Understanding and meaning do not come from methods and techniques. It seems that Western society is faced with a problem of understanding and meaning; not one of increased specialization. An understanding of the relationships between fields of specialization and an appreciation of the unities which underlie many fields of knowledge, are neither more or less difficult intellectually today than they were in Bacon's time. The important difference is that our society has placed its greatest tangible emphasis, its structure of rewards, and its social approval on specialization, while providing only secondary recognition and abstract approval to those who reject virtuosity alone and who argue for unity of conception and meaning removed from method.

The existence of this structuring of attitudes is demonstrated by our ease in accepting the caricatures of "literary brethren" as people who want to stop technology and ban phosphate wash powder; of "literary scholars" who are useful because they can tell us stories about the herculean feats of the past. Caricaturing is a dangerous art, because it allows us to dismiss the serious contributions of those caricatured with a laugh. Yet, many of these informed critics are attempting to argue that our current attitudes towards technology must be changed fundamentally, without giving up the genuine benefits which technology so obviously offers. Failure to communicate this critical goal lies at the heart of the problem of the two cultures. Neither the establishment of communication nor the transformation of attitude seems possible through the improvement of our systems engineering.

A different approach to the problem of two cultures is generated by the critical assessment of the myth of objectivity and the reliance upon specialty and technique. Perhaps, rather than emphasizing *fields* of specialization, we should emphasize the people having specialized talents. Perhaps we should distinguish between two types of people: those who are inclined and capable of understanding their own field of endeavor well enough to appreciate the interrelationships, between their own field and others, and those people who are only inclined or able to see their own field of specialized activity. Such a distinction would help to eliminate the problem of isolated disciplines at one level and would serve to delineate more clearly the reason for the isolation on another. Those with the capability and inclination to stress conceptual unities beyond the areas of their technical competence could be encouraged to elaborate and articulate the understandings which emerge from recognition of such unities. Creative artists—be they musicians, painters, scientists, or inventors—seldom have difficulty in communicating. They also consider technique and specialized training essential, but not sufficient, to their work. Emphasis upon all these men in all specialties, who endeavor to break their isolation from other people in other fields, would provide the basis for more informed critical evaluation of all forms of activity. Those with technical competence and no inclination to move past technique to understanding might be stimulated by this shift in emphasis to discover new perspectives. They might also be more clearly seen as people who represent the source of the problem of the two cultures. Such a transformation of attitude would mean, of course, sweeping changes in practices in our society. We would have to reorient, for example, our current approach to education and the way in which we perceive an educated person.

Changes of conceptualization in society are difficult and rare, but the current practices of our society have led to serious problems which will not rectify themselves. Change of action is brought about by change of attitude, value, and meaning. The demand for "immediate action" reveals a basic misconception. Presumably, the demand for action is a demand for meaningful, rather than haphazard, random activity. If this is the case, meaning must be discernible before an action takes place. The search for meaning, for understanding, and for unifying relationships is the foundation for meaningful action. We can no longer afford to act now and find meaning later, for our natural resources can no longer support such an approach. Neither, it seems, can the reservoir of our quality of life.

NOTES

(1) Both quotations appear in Gunther Steht, *The Coming of the Golden Age*, (New York: Doubleday, Natural History Press, 1969), pp. 73-74. He provides a stimulating discussion of the limits of molecular biology, as well as a discussion of what he considers to be an end point in Western society.

(2) Arthur Koestler, *The Act of Creation*, (New York: The MacMillan Company, 1964), p. 178.

(3) See Arthur Koestler, *The Roots of Coincidence*, American Edition, (New York: Random House, 1972) for a cautious and interesting general account of modern investigations of ESP and for highly suggestive ideas connecting the biological and physical sciences. He includes an excellent and easily accessible bibliography.

(4) See Arthur Koestler, *The Roots of Coincidence*, Chapter I, Section 8 for a description of these experiments; or see H. Schmidt, *Journal for Parapsychology*, Vols. XXXIII, No. 2, June 1969, No. 4, December 1969, XXXIV, No. 3, September 1970, No. 4 December 1970a and *New Scientist*, June 24, 1971.

(5) C. P. Snow, *The Two Cultures and A Second Look*, (Cambridge: Cambridge University Press, 1964)

BIBLIOGRAPHY

Baskin, Ted, *Quantum Theory and Beyond* (Cambridge University Press, 1971)

Burtt, A. E., *The Metaphysical Foundations of Modern Science* (Doubleday Anchor Books, 1954)

Good, I. J., editor, *The Scientist Speculates—An Anthology of Partly Baked Ideas* (London, 1962)

Koestler, Arthur, *The Act of Creation* (New York: The MacMillan Company, 1964)

Koestler, Arthur, *The Roots of Coincidence* (New York: Random House, 1972)

Koestler, Arthur, and J. R. Smythes, editors, *Beyond Reductionism—New Perspectives in the Life Sciences* (London, 1969)

Leonard, George B., *The Transformation* (New York: Delacorte Press, 1972)

Pearce, Joseph, *The Crack in the Cosmic Egg* (Pocket Book Edition, 1973)

Royce, Joseph R., *The Encapsulated Man* (VanNostrand Reinhold, An Insight Book, 1964)

Stent, Gunther, *The Coming of the Golden Age* (New York: The Natural History Press, 1969)

CONTRIBUTORS

Christopher W. Allen is an Associate Professor of Chemistry at the University of Vermont. After completing his undergraduate education at the University of Connecticut, he earned his M.S. and Ph.D. in Chemistry at the University of Illinois. His research has been in inorganic chemistry and he has published numerous papers on main-group inorganic and phosphorous chemistry.

A. Eugene Anderson, Director of Engineering at Western Electric's Allentown Works, in May, 1968, received the Distinguished Alumnus Award of Ohio State University from which he received the B.S. and Master of Science degrees, both in electrical engineering. After a career in the Bell Telephone Laboratories, which was interrupted by service in the Army Signal Corps Engineering Laboratories during World War II, he joined Western Electric in 1957. He holds a number of patents, is the author of several articles on transistors, and is active in a number of organizations, including the American Association for the Advancement of Science, Sigma Xi and Tau Beta Pi.

William E. Banton, Director of Manufacturing in Western Electric's Merrimack Valley Works in North Andover, Massachusetts, received his B.S. degree from Oregon State College, and his M.S. degree from Northeastern, both in electrical engineering and a M.S. degree in industrial management from MIT, which he attended as a Sloan Fellow. Before joining W.E. in 1951, he had been employed by the Monongahela Power Company and E.I. Dupont & Co. In Addition to his engineering experience, Mr. Banton has a background in systems development and management information.

Dr. Jerry Cassuto, General Medical Director of the Western Electric Company, received the Bachelor of Science degree cum laude in 1952 from the College of the City of New York and is a graduate of the State University College of Medicine. He also was awarded a Master of Science degree from the University of Rochester. Associated with Western Electric since 1964, he is active in the American Heart Association and the Industrial Medical Association.

Robert P. Clagett, an alumnus of the University of Maryland and a former Sloan Fellow at MIT, is General Manager of Material Planning and Merchandise in the Western Electric Company. The holder of several patents, he was one of the first development engineers assigned to the Company's Engineering Research Center.

Donald K. Conover, Director of Corporate Planning for Western Electric, received his Bachelor of Science degree from Princeton in 1953, and as a Sloan Fellow earned a Master of Science degree from MIT. An advisor to the Business Opportunities Workshop in Harlem, he is a member of the World Future Society and of the National Planning Association's Business Advisory Council on National Priorities.

George V. Cook at the time of the symposium was Western

Electric's Vice President in charge of regulatory matters. After graduation from high school in 1944, he enlisted in the U.S. Army as a private but before the completion of his service he attained the rank of first lieutenant in the Infantry. He received his B.A. from Columbia in 1949 and his LL.B. from the Columbia Law School in 1952. He practiced law with a large firm until 1956 when he joined the New York Telephone Company. He became associated with Western Electric in 1966. Mr. Cook is the author of a number of articles on legal subjects.

Cornelius L. Coyne, Jr., General Manager—Information Systems in Western Electric's Financial Division, has been associated with the Company since his graduation in 1952 from Illinois Benedictine College, except for a two-year period of service in the U.S. Army. He is the author of several articles on noise control.

Albert D. Crowell is Professor of Physics and Chairman of the Department of Physics at the University of Vermont. After receiving a B.S. at Brown University and an M.S. at Harvard University, he returned to Brown for a Ph.D. in physics. Professor Crowell has done extensive research in surface physics, especially on the interaction of gas molecules with solid surfaces. He has published many papers and is the co-author of the *Physical Adsorption of Gases on Solid Surfaces*.

John W. Engroff, Jr., is a doctoral candidate in the history of science at Harvard University, received a B.A. from Wesleyan University and an M.A. from the University of Wisconsin. Mr. Engroff is the Academic Coordinator of the University Year for Action at the University of Vermont, where he is also an Adjunct Assistant Professor of History and teaching courses in Islamic civilization.

Jeremy P. Felt is Professor of History and Chairman of the

History Department at the University of Vermont. After completing his undergraduate and master's work at Duke University, he earned a Ph.D. in history at Syracuse University. Professor Felt is the author of a book on child labor and has published many articles on American reform movements and the Progressive Era.

Stanley M. Grubin at the time of the symposium was General Manager of the Northeastern Division of Western Electric's Service Division. He began his career as an installer in March, 1945. While on a military leave of absence starting in February, 1952, he served in Korea and Japan. He returned to Western Electric as an installer, and was promoted to the position of job supervisor in 1956. He was on leave from January, 1959, to June, 1960, while attending Colorado University on a full-time basis. He received his B.S. in electrical engineering in 1960.

Raul Hilberg is Professor of Political Science and Chairman of the Department of Political Science at the University of Vermont. He did his undergraduate work at Brooklyn College and received an M.A. and a Ph.D. from Columbia University. A well-known scholar and lecturer, Professor Hilberg is the author of *The Destruction of the European Jews*.

Willard M. Miller did his undergraduate and master's work at the University of Illinois and received a Ph.D. in philosophy from the University of Rochester. An Assistant Professor of Philosophy at the University of Vermont, he is a student of the life and work of Charles S. Peirce, the noted American logician, philosopher, scientist and founder of pragmatism.

Henry John Steffens is an Associate Professor of History at the University of Vermont. He received his formal education at Cornell University, where he received a Ph.D. in the history of science. Professor Steffens' research interests include the history of the physical sciences in the eighteenth, nineteenth and twentieth centuries. His *A History of Newtonian Optics* is in press, and he is currently preparing a manuscript on the history of the conservation of energy principle in the nineteenth century.

L. Pearce Williams is the John Stambough Professor of History and Chairman of the Department of History at Cornell University. He received his Ph.D. from Cornell and taught at Yale University and the University of Delaware before joining the Cornell faculty. He is best known for his *Michael Faraday, A Biography* (1965) and for *Selected Correspondence of Michael Faraday* (1971), but his publications range broadly over the history of nineteenth century science.

A Note about This Book

The text of this book was set on an IBM Selectric Composer System in the type face Press Roman. The book was composed by Live Gold, Inc., New York and was printed and bound by Thomson-Shore, Dexter, Michigan. Production design by Robert Schroer. The jacket was designed by Donald Mowbray.